Principles of Environmental Science

Principles of Environmental Science

June Banks

SYRAWOOD
PUBLISHING HOUSE

New York

Published by Syrawood Publishing House,
750 Third Avenue, 9th Floor,
New York, NY 10017, USA
www.syrawoodpublishinghouse.com

Principles of Environmental Science
June Banks

International Standard Book Number: 978-1-64740-357-7 (Hardback)

This book contains information obtained from authentic and highly regarded sources. All chapters are published with permission under the Creative Commons Attribution Share Alike License or equivalent. A wide variety of references are listed. Permissions and sources are indicated; for detailed attributions, please refer to the permissions page. Reasonable efforts have been made to publish reliable data and information, but the authors, editors and publisher cannot assume any responsibility for the validity of all materials or the consequences of their use.

Trademark Notice: Registered trademark of products or corporate names are used only for explanation and identification without intent to infringe.

Cataloging-in-Publication Data

Principles of environmental science / June Banks.
   p. cm.
Includes bibliographical references and index.
ISBN 978-1-64740-357-7
1. Environmental sciences. 2. Ecology. I. Banks, June.
GE105 .H36 2023
363.7--dc23

TABLE OF CONTENTS

Permissions

Index

PREFACE

Environmental science is an interdisciplinary field of study that integrates physics, biology, geography, ecology, geology, mineralogy and chemistry to examine environmental problems along with their impact on humans, and providing solutions to these environmental problems. Environmental studies is another field which is closely associated with environmental science, and it incorporates social sciences to understand the human relationships, perceptions, and policies towards the environment. Environmental science seeks to comprehend the physical, chemical, biological, and geological processes of the Earth. This knowledge is used to understand the different phenomena and solutions such as alternative energy systems, pollution control and mitigation, natural resource management, and the effects of global warming and climate change on the natural systems and processes of the Earth. The objective of this book is to provide an overview of the principles of environmental science. It will prove to be immensely beneficial to students and researchers in this field.

All of the data presented henceforth, was collaborated in the wake of recent advancements in the field. The aim of this book is to present the diversified developments from across the globe in a comprehensible manner. The opinions expressed in each chapter belong solely to the contributing authors. Their interpretations of the topics are the integral part of this book, which I have carefully compiled for a better understanding of the readers.

At the end, I would like to thank all those who dedicated their time and efforts for the successful completion of this book. I also wish to convey my gratitude towards my friends and family who supported me at every step.

June Banks

Chapter 1

Introduction to Environment

1.1 Environmental Studies: Meaning, Scope and Importance

Definitions of Environment

Environment is defined as the surroundings or conditions in which a person, animal or plant lives or operates.

Some important definitions of environment given by well-known environmentalist are given below:

1. Boring

An environment consists of the sum of the stimulation which receives from conception until death. It can be concluded from the above definition that, "Environment comprises of various types of forces such as physical, political, intellectual and economic, social, cultural, moral and emotional".

2. Douglas and Holland

The term environment is used to describe all the external forces, influences and conditions, which affect the life, nature, behaviour, growth, development and maturity of living organisms.

Meaning of Environmental Study

Environmental study deals with the analysis of the processes in air, water, soil, land and organisms which leads to pollute or degrade the environment.

It helps us for establishing the standard for safe, clean and healthy natural ecosystem. It also deals with important issues like safe and clean drinking water, fertility of land, hygienic living conditions and clean and fresh air, healthy food and development.

The environmental studies includes not only the study of physical and biological characters of the atmosphere but also the social and cultural factors and the impact of man on environment.

Environmental property is the ability to keep up the rates of natural resources harvest, pollution creation and non-renewable resource depletion that may be continued indefinitely.

Scope of Environmental Studies

The scope of this study is listed below:

- Natural resources - for their conservation and management.

- Ecology and biodiversity.

- Environmental pollution and control.

- Social issues in relation to development and environment.

- Human population and environment.

The above are the basic aspects of environmental studies which have a direct relevance to every section of the society.

Importance of Environmental Studies

The environmental studies enlighten us, about the importance of conservation and protection of our environment from indiscriminate release of pollution. At present a great number of environmental issues have grown by size and complexity, threatening the survival of mankind on earth.

Environmental studies have become significant for the following reasons:

1. Environmental Issues Being of International Importance

It has been well recognized that environmental issues such as acid rain, biodiversity, global warming, marine pollution and ozone depletion are not merely national issues but these are global issues and hence it must be tackled with international efforts and co-operation.

2. Problems Cropped in the Wake of Development

Development, in its wake gave birth to agriculture and housing, industrial growth, transportation systems, urbanization, etc. However, it has become phased out in the developed world. The North, to cleanse their own environment has fact fully managed to move 'dirty' factories on South.

When the West developed, it did so perhaps neglecting the environmental impact of its activities. Evidently such a path is neither desirable nor practicable, even if developing countries follows that.

3. Explosively Increase in Population

World census reflects that one in every seven person in this planet lives in India. Evidently with 16% of the world's population and only 2.4% of its land area, there is a heavy pressure on the natural resources including land.

Agricultural experts have recognized soil health problems such as deficiency of micronutrients and organic matter, soil salinity and damage of soil structure.

4. Need for an Alternative Solution

It is essential especially for developing countries to find alternative paths to attain alternative goals.

5. Need to Save Humanity from Extinction

It is an our responsibility to save the humanity from extinction. As a result of development, the environment has been constricted and the biosphere has been depleted.

6. Need for Wise Planning of Development

Our survival and sustenance depend. Resources withdraw, processing and use of the products have to be synchronized with the ecological cycles in any plan of development, our actions should be planned ecologically for the sustenance of the environment and development.

1.1.1 Need for Public Awareness

Man has acquired the capacity to change the environment more than any other organism on this planet. He uses much more material and energy for his agriculture, industry, transport, comfort, communication, aesthetic pleasure and even war than any other species on the earth.

Human needs and greeds have disturbed the delicate ecological balance. Humans are depleting and degrading the vital life supporting systems including air, water and land to the entire living world. Any government at its own level cannot achieve the goals of sustainable development until the public has a participatory role in it. Public participation is possible only when the public is aware about the ecological and environmental issues.

There are several laws enacted by the Government of India for conservation and prevention of pollution. But law cannot be implemented unless education makes society aware of the risks of living in a deteriorating environment.

Education provides the people with adequate knowledge to fully analyse the environmental problems and take appropriate actions. It use our talents creatively to improve the quality of our environment.

The UN conference on Human Environment held in Stockholm from 5th-6th June, 1972, which had the distinction of the Prime Minister of India, late Indira Gandhi addressing it, was a major breakthrough in creating awareness. From then on the World Environment Day is observed every year on June 5 with a view of arousing the masses and bringing environmental issues to the forefront of thinking and planning.

The controversy over the Silent Valley hydel project to save "one of the last vestiges" of tropical rain forest, the "Chipko-Movement" in Himalayas (Almora Hills) and the Appiko movement (Karnataka) in India indicate people's awareness.

Environmental problems can often seem intractable and the change necessary to solve them elusive. But we have no choice but to change. To fail to change is to deny the responsibility we have for one another and for those who come after us.

Young people who will be among the leaders of the new generation of environmental protection and environmental leaden they must build environmental democracy. An informed local community always does a better job of environmental protection than a distant bureaucracy.

1.2   Segments of Environment

The environment consists of four segments. They are described below:

1. Atmosphere

The atmosphere implies the protective blanket of gases, surrounding the earth:

- It sustains life of Earth.
- It saves from the hostile environment of outer space.
- It absorbs most of the cosmic rays from the outer space and a major portion of the electromagnetic radiation from the sun.
- It transmits only near infrared radiation, radio waves, ultraviolet and visible waves. While filtering out tissue damaging ultraviolet waves below 300mm.

The atmosphere is composed of nitrogen and oxygen. Besides, argon, carbon dioxide and trace gases.

2. Hydrosphere

The Hydrosphere comprises of all types of water resources such as glaciers, ground water, lakes, oceans, polar icecaps, reservoir, rivers, seas and streams:

- 97% of the earth's water supply is ocean.

- About 2% of the water resources are locked in either polar icecaps or glaciers.

- Only about 1% is available as fresh surface water such as ground water, lakes, rivers, streams and fit to be used for human consumption and other uses.

3. Lithosphere

Lithosphere is an outer mantle of the solid earth. It comprises of minerals occurring in the earth's crusts and the soil. Some examples are air, minerals or ganic matter and water.

4. Biosphere

Biosphere indicates the realm of living organisms and their interactions with the environment, viz lithosphere, hydrosphere and atmosphere.

1.2.1 Environmental Degradation

It is the process by which our environment i.e., air, water and land, is progressively contaminated, over-exploited and destroyed. When the environment becomes less valuable or damaged, environmental degradation is said to occur.

In specific term, Environmental degradation is the disintegration of the earth or deterioration of the environment through consumption of assets, for example, air, water and soil; the destruction of environments and the eradication of wildlife. It is characterized by any change or aggravation to nature's turf seen to be pernicious or undesirable.

Ecological effect or degradation is created by the consolidation of an effectively substantial and expanding human populace, constantly expanding monetary development or per capita fortune and the application of asset exhausting and polluting technology.

It occurs when earth's natural resources are depleted and the environment is compromised in the form of extinction of species, pollution of air, water and soil and rapid growth in population.

Causes of Environmental Degradation

Based on the discussion so far, we now know that healthy environment is essential for the very existence of human society and other living organisms. But environmental degradation is going on unabated.

We are being cautioned every now and then about the deterioration in the environment and its consequences like global warming, changing climatic conditions, impending water crisis, decreasing fertility of agricultural land and increasing health problems. There is an urgent need to take all possible steps to check environmental degradation.

In order to consider the required steps to be taken for doing so, it is necessary to understand the causes of environmental degradation. The important factors are the following:

1. Growing Population

A population of over thousands of millions is growing at 2.11 per cent every year. It puts considerable pressure on its natural resources and reduces the gains of development. Hence, the greatest challenge before us is to limit the population growth. Although population control does automatically lead to development, yet the development leads to a decrease in population growth rates.

2. Poverty

India has often been described a rich land with poor people. The poverty and environmental degradation have a nexus between them. The vast majority of our people are directly dependent on the natural resources of the country for their basic needs of food, fuel shelter and fodder. About 40% of our people are still below the poverty line.

Environment degradation has adversely affected the poor who depend upon the resources of their immediate surroundings. Thus, the challenge of poverty and the challenging environment degradation are two facts of the same challenge. The population growth is essentially a function of poverty. Because, to the very poor, every child is an earner and helper and global concerns have little relevance for him.

3. Agricultural Growth

The people must be acquainted with the methods to sustain and increase agricultural growth with damaging the environment. High yielding varieties have caused soil salinity and damage to the physical structure of the soil.

4. Need to Ground Water

It is essential of rationalizing the use of groundwater. Factors like community wastes, industrial effluents and chemical fertilizers and pesticides have polluted our surface water and affected quality of the groundwater. It is essential to restore the water quality of our rivers and another water body as lakes is an important challenge.

It so finding our suitable strategies for the consecration of water, provision of safe drinking water and keeping water bodies clean which are difficult challenges is essential.

5. Development and Forests

Forests serve catchments for the rivers. With increasing demand of water, plan to harness the mighty river through large irrigation projects were made. Certainly, these would submerge forests; displace local people, damage flora and fauna.

Forests in India have been shrinking for several centuries owing to pressures of agriculture and other uses. Vast areas that were once green, stand today as wastelands. These areas are to be brought back under vegetative cover.

The tribal communities inhabiting forests respect the trees and birds and animal that gives them sustenance. We must recognize the role of these people in restoring and conserving forests.

The modern knowledge and skills of the forest department should be integrated with the traditional knowledge and experience of the local communities. The strategies for the joint management of forests should be evolved in a well-planned way.

6. Reorientation of Institutions

The people should be roused to orient institutions, attitudes and infrastructures, to suit conditions and needs today. The change has to be brought in keeping in view India's traditions for resources use managements and education etc.

Change should be brought in education, in attitudes, in administrative procedures and in institutions. Because it affects way people view technology resources and development.

7. Reduction of Genetic Diversity

At present most wild genetic stocks have been disappearing from nature. Wilding including the Asiatic Lion is facing the problem of loss of genetic diversity. The protected areas network like sanctuaries, national parks, biosphere reserves is isolating populations. So, they are decreasing changes of one group breeding with another. Remedial steps are to be taken to check decreasing genetic diversity.

8. Land Disturbance

A more basic cause of environmental degradation is land damage. Numerous weedy plant species, for example, garlic mustard, are both foreign and obtrusive. A rupture in the environmental surroundings provides for them a chance to start growing and spreading. These plants can assume control over nature, eliminating the local greenery. The result is the territory with a solitary predominant plant which doesn't give satisfactory food assets to all the environmental life. Whole environments can be destroyed because of these invasive species.

9. Pollution

Pollution, in whatever form, whether it is air, water, land or noise is harmful to the environment. Air pollution pollutes the air that we breathe which causes health issues. Water pollution degrades the quality of water that we use for drinking purposes. Land pollution results in degradation of earth's surface as a result of human activities. Noise pollution can cause irreparable damage to our ears when exposed to continuous large sounds like honking of vehicles on a busy road or machines producing large noise in a factory or a mill.

10. Overpopulation

Rapid population growth puts the strain on natural resources which result in degradation of our environment. The mortality rate has gone down due to better medical facilities which have resulted in increased lifespan. More population simple means more demand for food, clothes and shelter. We need more space to grow food and provide homes to millions of people. This results in deforestation which is another factor of environmental degradation.

11. Landfills

Landfills pollute the environment and destroy the beauty of the city. Landfills come within the city due to the large amount of waste that gets generated by households, industries, factories and hospitals. Landfills pose a great risk to the health of the environment and the people who live there. Landfills produce a foul smell when burned and cause huge environmental degradation.

12. Natural Causes

Things like avalanches, quakes, tidal waves, storms and wildfires can totally crush nearby animal and plant groups to the point where they can no longer survive in those areas. This can either come to fruition through physical demolition as the result of a specific disaster or by the long term degradation of assets by the presentation of an obtrusive foreign species to the environment. The latter frequently happens after tidal waves, when reptiles and bugs are washed ashore.

Effects of Environmental Degradation

1. Impact on Human Health

Human health might be at the receiving end as a result of the environmental degradation. Areas exposed to toxic air pollutants can cause respiratory problems like pneumonia and asthma. Millions of people are known to have died of due to indirect effects of air pollution.

2. Loss of Biodiversity

Biodiversity is important for maintaining balance of the ecosystem in the form of combating pollution, restoring nutrients, protecting water sources and stabilizing climate. Deforestation, global warming, overpopulation and pollution are few of the major causes for loss of biodiversity.

3. Ozone Layer Depletion

Ozone layer is responsible for protecting earth from harmful ultraviolet rays. The presence of chlorofluorocarbons, hydro-chlorofluorocarbons in the atmosphere is causing the ozone layer to deplete. As it will deplete, it will emit harmful radiations back to the earth.

4. Loss for Tourism Industry

The deterioration of environment can be a huge setback for tourism industry that relies on tourists for their daily livelihood. Environmental damage in the form of loss of green cover, loss of biodiversity, huge landfills, increased air and water pollution can be a big turn off for most of the tourists.

5. Economic Impact

The huge cost that a country may have to borne due to environmental degradation can have big economic impact in terms of restoration of green cover, cleaning up of landfills and protection of endangered species. The economic impact can also be in terms of loss of tourism industry.

1.3   Role of Individual in Environmental Conservation

Resources are being exhaustible, it is the duty of every individual on this earth to conserve the natural resources in such a way that they must be available for future generation also. Individual must understand the essential of natural resources. Due to advancement in technology and population growth, the present world is feeling lot of problems on degradation of natural resources.

Measures recommended for conservation of natural resources:

1. Conservation of Energy

- Switch off lights, fans and other appliances when not in use.
- Use solar heater for cooking our food on sunny days, which will cut down our LPG expenses.
- Dry the clothes in sunlight instead of driers.

- Grow trees near the houses and get a cool breeze and shade. These will cut-off our electricity charges on A/C and waters.

- Use always pressure cooker.

- Ride bicycle or just walk instead of using car and scooter.

2. Conservation of Water

- Use minimum water for all domestic purposes.

- Check for water leaks in pipes and toilet and repair them promptly.

- Use drip irrigation to improve the irrigation efficiency and reduce evaporation.

- Reuse the soapy water, after washing clothes, for washing off the country yards, drive ways, etc.

- Build rainwater harvesting system in our house.

- The wasted water, coming out from kitchen, bath tub, can be used for watering the plants.

3. Conservation of Soil

- While constructing the house don't uproot the trees as for as possible.

- Grow different types of plants, herbs, trees and grass in our garden and open areas, which bind the soil and prevent its erosion.

- Soil erosion can be prevented by the use of sprinkling irrigation.

- Don't irrigate the plants using a strong flow of water, as it will wash off the top soil.

- Uses mixed cropping, so that some specific soil nutrients will not get depleted.

- Use green manure in the garden, which will protect the soil.

4. Conservation of Food Resources

- Eat only minimum amount of food. Avoid over eating.

- Don't wastes the foods instead give it to someone before getting spoiled.

- Cook only required amount of the food.

- Don't cook food unnecessarily.

- Don't store large amounts of food grains and protect them from damaging insects.

5. Conservation of Forest

- Use non-timer products.

- Plant more trees and protect them.

- Grassing, fishing must be controlled.

- Minimize the use of papers and fuel wood.

- Avoid execution of developmental work like dam, road, construction in forest areas.

1.3.1 Sustainable Lifestyle

The life support system of the biosphere is constantly threatened due to degradation of natural resources and expanding population. Rise in technological activities are contributing a lot to rapid and stressful changes in our global environment.

Agriculture, forestry, various land use practice, industrial activities and waste disposal have seriously altered the natural ecosystem, affecting the biological productivity and the chemistry of the atmosphere. The environmental problems would influence the developing nations, mostly due to lack of practicing conservation norms resulting in the loss of biodiversity, deforestation, alternation of earth's climate and desertification.

The deteriorating state of global environment and increasing human demand on resources have lead to concern about the sustainability of earth's life support base.

The four basis causes of environmental imbalances are:

- Rapid increase in population.

- Pollution.

- Increased consumption.

- Land deterioration.

The term sustainable means something that conserves ecological balance by avoiding depletion of national resources.

The natural resources may be renewable and non-renewable. Renewable resources are incxhaustiblc and infinite.

Renewable resources are also called biotic resources and they include agriculture, forestry and fishery products.

Non-renewable resources are exhaustible and finite stock resources. They are also called abiotic and they include minerals, metals, soils, coal, oil and natural gas. There is a continuous flow and consumption of energy in the ecosystem functioning and technological civilization.

This has caused rapid decline in the quality and concern for their conservation and management. Thus concern for conservation started.

Practically speaking conservation refers to the planning and management of natural resources, to secure their wise use, continuity of their supply, maintaining their quality and chancing upon the exploitation of natural resources.

Among the Indian natural resources India is number 1 in coal, it also stands at 3rd for Mn and 4th for Fe. At present India is facing ecological crisis that's why the natural resources are being declined/depleted. Specially the resource needed for human survival and sustainable development, however are being depleted rapidly.

Keeping in view this problem of depletion of natural resources, world conservation strategy was launched in march 1988 for guidance and objectives for conservational of natural resources at national and international level.

It had the following objectives:

- Living resources conservation is a limited sector and must be considered by all sectors.

- Consequent failure to integrate conservation with development.

- Destructive development process due to inadequate environment planning and management.

- Lack of capacity to conserve due to lack of legislation and its enforcement.

- Lack of awareness to support conservation.

- Lack of conservation in rural area of developing countries.

Certain requirements have been laid down by world conservation strategy (WCS) to achieve conservation:

- Conservation should be incorporated in conservation and law should enforce it.

- Specific legislation for conservation by providing both sustainable utilization and protection of living resources.

- Ecological consideration in policy making.

- Public education and conservation.

- Monitoring of effectiveness of law.

The environmental problems that challenge human, society are ecological in nature. Therefore, there is an urgent need to address economical activity, population growth and environmental protection as interrelated issues.

The Rio conference on environment and development in 1992 emphasized a close link between environmental problems and economic development. The concept of growth at any cost has to be replaced with that of sustainable development that implies equitable distribution of resources in harmony with nature.

The local environment and ecological system are to be used in a manner that satisfies the current need of options without adjustments with the welfare of the future generations. The "Agenda 21" of UN Convention is a comprehensive action plan for moving human kind into the age of sustainability.

The task of achieving sustainable development while conserving the resources is a daunting one, but it could be achieved through environment friendly decentralized planning, scientific evaluation of resources, clean production technologies, propagation of indigenous knowledge using local resources and community co-operation.

Chapter 2

Natural Resources

2.1   Forest Resources

Forest, an important resource to mankind, supports in a direct or indirect way the development of human economy and society with its thousands of material or non-material products. The forest is also the main object for the continental ecological system.

The successful management of forest inheritance plays a vital role in managing the global environment, conserving the other related resources and guaranteeing the sustainable development of economy and society.

Uses of Forest Resources

The uses or functions of forest resources can be classified into three categories are:

- Protective uses.

- Regulatory uses.

- Productive uses.

Protective uses

- Trees provide excellent protection to the soil against soil erosion. The deep and spreading roots of trees and other plants in the forest hold the soil particles strongly and prevent them from being washed away in rain.

- Forest trees build up a thick layer of humus to prevent floods, droughts and soil erosion.

- Roots of trees and other plants serve as a natural dam and hold rain water.

- Forests prevent early siltation of dams and lakes.

In the absence of trees, soil at the edges fall continuously into the water bodies making them shallow, reducing their water storage capacity. This is especially bad for dams because when a dam is built, its life is usually predicted. Siltation shortens the life span of dams. Siltation also increases floods.

Regulatory uses

- Forests regulate various biogeochemical cycles, especially that of carbon. This is because, forest trees absorb huge quantities of carbon dioxide from the air during photosynthesis.

- During photosynthesis, trees release huge quantities of life giving oxygen, which is one of the end products of the process.

- Trees help purify the atmosphere. Thereby reducing pollution drastically.

- Acting as sponges, forests absorb and hold water which charges springs, streams and ground water. Thus they regulate the flow of water from mountain highlands to croplands and urban areas.

- Forests also influence local, regional and global climate.

- Trees absorb large quantities of water from the soil and release it into the atmosphere during the process of transpiration. About 50 to 80% of moisture in the air above tropical forests comes from trees in this manner. This moisture soon becomes rain clouds. Therefore, if large areas of forests are cleared, rainfall will be affected.

- Wild animals and plants are most essential to balance the nature itself. In their absence, ecosystem will be misbalanced.

- Forests buffer us against noise, absorb air pollutants and raise human spirits.

Productive uses

Worldwide, about half the timber cut each year is used as fuel for heating and cooking, especially in the poor countries. Some is burnt directly as firewood, while the rest is converted to charcoal, which is used in urban areas in poor countries and also by many industries.

Over Exploitation of Forests

Most of the overexploitation of forests is taking place in the tropical countries. Tropical forests are of different types and cover about 6% of Earth's land area. All types of forests are affected, but the worst situation is in the rainforests.

Forests are Overexploited in Different Manners

The most important methods of overexploitation are deforestation, mining and dam building. It is often thought that tribal people destroy the forests in which they live. However, this is not true. The quantum of damage done by, the forest dweller is not much. They depend on the forests and their produce for livelihood. The real causes lie elsewhere.

Deforestation, Mining and Dams and their Effects on Forests and the Tribal People The three main reasons for the dwindling forest resources in the world are defor-estation, mining and dam building. When the area under forest cover is decreased due to any reason, it is the overall prosperity of the country that suffers. And the people most affected are the tribal people who depend on the forests for their live-lihood.

The tribal people live a simple life. They get most of their food in the form of leaves, fruits, roots. rhizomes. etc. and the small animals from the forests. They drink water from the streams of the forests. They collect firewood for cooking and also for selling in the nearby villages. They also sell flowers, fruits, leaves, twigs for brushing teeth, hon-ey, lac, gum, fibres and other such forest produce to earn livelihood.

They make huts using bamboos and other building materials available in the forests. They use the medicinal plants available in the forests for different types of ailments. In this way, the life of the forest dwellers is dependent entirely on the forests. Therefore, the direct effect of deforestation by any method whether it be lumbering or mining or dam building is felt by the tribal people the most, though everybody else is affected too in various ways.

Reason for Over Exploration in India

It has been estimated that in India the minimum density of forests required to maintain good ecological balance is about 33, all area. But, at present it is only about 22%. So over exploitation forest materials occur.

Causes of Over Exploitation

Over exploitation of forest wealth in the developing countries occurs in the following ways:

- Increasing agricultural production.

- Increasing industrial activities.

- Increasing in demand of wood resources.

2.1.1 Deforestation, Mining, Dams and their Effects on Forest and Tribal people

The removal of trees from forest or woodland areas is currently a major environmental problem of the developing countries, most of which are in the tropics. Tropical forests account for nearly half of the world's remaining forests:

- Deforestation is caused by a number of human activities. For example, the increase in the practice of shifting agriculture is a primary cause. This type of extensive agriculture has coexisted with tropical rain forest for thousands of years.

- Other agricultural practices responsible for deforestation are the establishment of plantations for cash crops and the use of cleared forest areas for cattle ranching. Other causes of deforestation are mineral extraction, the building of roads and dams and logging operations.

- There are a number of environmental problems associated with deforestation. On a local level, the protective tree cover plays a very important role in the water cycle. The water taken up by the trees reduces the amount available for surface runoff. When deforestation occurs, surface runoff increases.

- This situation can lead to severe problems of accelerated soil erosion. For example, in mountainous regions of Nepal and northern India, entire hillsides, stripped of their natural tree cover, have been washed away by heavy monsoon rains.

- On a regional level, the increase in surface runoff associated with deforestation has caused severe downstream flooding, while the transport of large amounts of sediment can destroy neighbouring areas and also cause problems of siltation.

- On a global scale, the burning of biomass associated with deforestation is contributing to an increase in atmospheric carbon dioxide levels. Large amounts of carbon dioxide are taken up by the forests during photosynthesis and extensive deforestation has resulted in the contraction of this important sink mechanism. Deforestation is a major contributor to the decline in global biodiversity.

- Deforestation has an important influence on regional and global climate. Deforestation affects regional climate by altering sensible and latent heat. Where deforestation has eliminated plants and animals and degraded water supplies and soil fertility. Major deforestation can cause the displacement of whole communities.

Deforestation is the process of elimination (or) removal of forest resources due to many natural or man-made activities. In general, deforestation means destruction of forests.

Causes of Deforestation

Developmental Projects:

Developmental projects cause deforestation in two ways;

- Through submergence of forest area under water.

- Destruction of forest area.

Examples: Hydroelectric projects, big dams, road constructions, etc.

Hence, there is a need to discourage the undertaking of any development works in the forest area.

Mining Operations:

Mining have a serious impact on forest areas. Mining operation reduces the forest area.

Example: Mica, Coal, Limestone, etc.

Raw Materials for Industries:

Wood is the important raw material for so many purposes.

Example: For making boxes, ply woods, etc.

Fuel Requirements:

In India, both tribal and rural population is dependent on the forest for meeting their daily needs of fuel wood, which leads to the pressures on forest, ultimately to deforestation.

Shifting Cultivation:

The replacement of natural forest ecosystem for mono specific tree plantation can lead to disappearance of number of plant and animal species.

Forest Fires:

Forest fire is one of the major causes for deforestation. Due to human interruption and rise in ambient temperature.

Effects of Deforestation on the Investment

Since many people are dependent on the forest resources, deforestation will have the following social, ecological and economic effects:

1. Soil Erosion

Deforestation also causes soil erosion, drought, landslides, floods. Natural vegetation acts as a natural barrier to reduce the wind velocity, which in turn reduces soil erosion.

2. Loss of Biodiversity

Most of the species are very sensitive to any disturbance and changes.

When the plants no longer exist, animals depend on them for food and habitat become extinct.

3. Loss of Food Grains

As a result of soil erosion, the countries loose the food grains.

Mining

Mining from shallow deposits is done by surface mining while that from deep deposits is done by sub-surface mining. It leads to degradation of lands and loss of top soil. It is estimated that about eighty thousands hectare land is under stress of mining activities in India.

Mining leads to drying up perennial sources of water sources like streams and spring in mountainous area. Mining and other associated activities remove vegetation along with underlying soil mantle, which results in destruction of topography and landscape in the area. Large scale deforestation has been reported in Mussorie and Dehradun valley due to indiscriminating mining.

The forested area has declined at an average rate of about 33% and the increase in non-forest area due to mining activities has resulted in relatively unstable zones leading to landslides.

Indiscriminate mining in forests of Goa since 1961 has destroyed more than 50000ha of forest land. The coal mining in Jharia, Raniganj and Singrauli areas has caused extensive deforestation in Jharkhand.

Mining of magnetite and soapstone have destroyed 14ha of forest in hilly slopes of Khirakot, Kosi valley and Almora.

Mining of the radioactive minerals in Kerala, Tamilnadu and Karnataka are posing similar threats of deforestation.

The rich forests of Western Ghats are also facing the same threat due to mining projects for excavation of copper, chromites, bauxite and magnetite.

Dams and other Effects on Forest and Tribal people

Pandit Jawaharlal Nehru referred dam and valley projects as "Temples of modern India". These big dams and rivers valley projects have multipurpose uses. However, these dams are also responsible for the destruction of forests. They are responsible for loss of flora and fauna, degradation of catchment areas, disturbance in forest ecosystems, and increase of water borne diseases, rehabilitation and resettlement of tribal peoples.

India has more than 1550 large dams, the maximum being in the state of Maharashtra, followed by Gujarat and Madhya Pradesh.

The highest one is Tehri dam, on river Bhagirathi in Uttaranchal and the largest in terms of capacity is Bhakra dam on river Satluj in Himachal Pradesh. Big dams have been in sharp focus of various environmental groups all over world, which is mainly because of several ecological problems including deforestation and socioeconomic problems related to tribal or native people associated with them.

The Silent valley hydroelectric project was one of the first such projects situated in the tropical rain forest area of Western Ghats which attracted much concern of the people.

The crusade against ecological damage and deforestation caused due to Tehri dam was led by Shri. Sunder Lal Bahaguna, the leader of Chipko Movement.

The cause of Sardar Sarovar Dam related issues have been taken up by the environmental activitist Medha Patkar, joined by Arundhati Ray and Baba Amte. For building big dams, large scale devastation of forests takes place which breaks the natural ecological balance of the region.

Droughts, floods and landslides become more prevalent in such areas. Forests are the repositories of invaluable gifts of nature in the form of biodiversity and by destroying them, we are going to lose these species even before knowing them.

These species could be having marvellous economic or medicinal value and deforestation results in loss of this storehouse of species which have evolved over millions of years in a single stroke.

Effects of Dams

Dams are massive artificial structures built across the river to create a reservoir in order to store water for many beneficial purposes. However, these dams are also responsible for the destruction of forest and displacement of local people.

The Indian Scenario

India has more than 1600 large dams. Dam is the highest built across the river in the State.

State	No. of Dams
Maharashtra	More than 500 dams
Gujarat	More than 250 dams
Pradesh	More than 130 dams

Effects of Dam on Forest

- Thousands of hectares of forest have been cleared for executing river valley projects.

- In addition to the dam construction, the forest is also cleared for residential accommodation, office buildings, etc.

- The big river valley projects also cause water logging which leads to salinity and in turn reduces the fertility of the land.

- Hydroelectric projects also have led to widespread loss of forest in recent years.

- Hydroelectric projects provide opportunities for the spread of water borne diseases.

- Construction of dams under these projects lead to killing of wild animals and destroying aquatic life.

Examples:

- Narmada Sugar Project: It has submerged 3.5 each hectares of forest comprising of teak and bamboo trees.

- Dam: It has submerged 1000 hectares of forest affecting about 430 species of plants.

Effects of Dam on Tribal people

- The greatest social cost of big dam is the widespread displacement of tribal people such that biodiversity cannot be tolerated.

- The displacement and cultural change affects the tribal people both mentally and physically. They do not accommodate modern food habits and life styles.

- Tribal people and their culture cannot be questioned and destroyed.

- Generally, the body conditions of the tribal people will not suit with the new areas and hence they will be affected by many diseases.

Conflicts of Water Resources

1. Conflicts Through use

Unequal distribution of water led to interstate or international disputes.

- International conflicts: India and Pakistan fight to get water from the Indus.

2. Construction of Dams/Power Stations

For hydroelectric power generation, dams built across the rivers, initiates conflict between the states.

3. Conflict Through Pollution

- Rivers and Lakes are used for electricity, shipping and for industrial purpose.

- Disposal of waste water and industrial waste decrease the quality of water and causes pollution.

2.2   Water Resources

The water cycle, through evaporation and precipitation, maintains hydrological systems which form rivers and lakes and support in a variety of aquatic ecosystems.

The wetlands are intermediate forms between the terrestrial and aquatic ecosystems and contain species of plants and animals that are highly moisture dependent. The aquatic ecosystems are used by several people for their daily needs such as drinking water, cooking, washing, watering animals and irrigating fields.

The world depends on a limited quantity of fresh water. Water covers 70% of the surface of earth but only 3% of this is fresh water of this, 2% is in polar ice caps and the remaining 1% is usable water in rivers, lakes and subsoil aquifers. Only a fraction of this can be actually used.

At a global level, 70% of water is used for agriculture about 25% for industry and only 5% for domestic use. Although this may varies in different countries, industrialized countries use a greater percentage for industry.

India uses 90% for agriculture, 7% for industry and 3% for domestic use. One of the greatest challenges facing the world in this century is the need to rethink the overall management of water resources.

Use and Over-Utilization of Water

It transpires from our water budget that, in case, average annual rainfall of entire country and its total area are taken, the total water resources are of the order of 167 mha m.

In fact, only 66 mha m of water can be utilized by us for irrigation. As there are sonic financial and technological constraints, we plan to use it fully only by AD 2010. By 1951, only 9.7 million m of water was used for irrigation. By 1973, it was as much as 18.4 mha m.

It is observed that agriculture sector is the major user of water. The water used for irrigation two decades was back which was nearly 40 per cent, has gone up to 73% by AD 2000. Irrigation use is very inefficient. Hence, 25-30percent efficiency and methods of irrigation are to be changed drastically. From the data on water use shown in the table.

Water use (India) AD 2000 (available water 1,900 million m3 per year.

Use	Taken	Consumed	Returned
Irrigation and livestock	869	783	86
Power	150	5	145
Industry	35	10	25
Domestic	38	8	30
Total	1,092	806	286

It becomes evident that irrigation inducing for livestock and power use is 79.6% and 13.7% water respectively. Thereafter, comes domestic (3.5 percent) and industrial (3.3 percent) uses. In case the land area is taken up as a unit, the position could be different. By 1984-85, the land under irrigation almost tripled to 67.5 mha.

After a period of five years, that is, by 1990 another 13 mha were to be brought under irrigation, thus, the total figure was 80 mha. This may be adjudged against the total potential of 133 mha by AD 2010. Here it may be kept in mind that it is the gross sown area and not the net sown area. The former, that is, the gross sown area is bound to be larger than the latter, that is, net sown area. At present, more than 30percent of the net sown area is under irrigation.

It is estimated by the World Health Organization (WHO) that water consumption will have to be cut by 50% by 2025 if nations fail to redress imbalances in global water supply and demand.

Some European companies have begun to supply water to water-thirsty countries across the oceans. Nordic Water Supply (Norwegian company) has been transporting freshwater, that is, clean drinking water in giant floating bags across the oceans.

These floating bags are sausage-shaped, about 200 m long. Each contains 35,000 tonnes of water. The floating water bags are made of a polyester fabric coated with plastic and are 2.0mm. (0.08.in.) thick. In future, the company plans to build new bags of the size of supertanker, 300m long and a capacity of 100,000 tons of water.

In this way, the Nordic company will be engaged in the business of towing freshwater from Turkey to Greek island. Its future plans include transporting water from Iran to Saudi Arabia along the Caribbean Sea and the Red Sea.

Ground Water

The replenish able groundwater potential in India is estimated at 433.9 billion cubic metres. Water percolates easily in the alluvial soils and hence the potential of the groundwater development is high in the Great Plains of India.

Uttar Pradesh alone accounts for 19.0% of the estimated groundwater potential. More than 42percent of the potential is confined to States of the Great Plains of north India.

Contrary to it, seepage of water in the rocky lands of peninsular India is slow, resulting in low groundwater potential. However, because of their size Maharashtra, Madhya Pradesh and Tamil Nadu also have large potential of groundwater resources of the total groundwater resources one-fourth is used for domestic, industrial and related purposes and three-fourths for irrigation.

Only 37.23% of the total available groundwater resources have so far been developed in India. State-wise percentage of developed groundwater resources to total available potential ranges from 1.07% in Jammu and Kashmir to 98.34 percent in Punjab.

Those States and union territories where there is scarcity of surface water due to low and highly variable rainfall have developed their groundwater resources on large scale.

Punjab, Haryana, western Uttar Pradesh, Rajasthan, Gujarat and Tamil Nadu are such States. There is need for the development of groundwater resources in Andhra Pradesh, Madhya Pradesh, Chhattisgarh, Karnataka and Maharashtra also where rainfall is comparatively insufficient and variable.

Over use of Underground Water

The water table has been falling rapidly in many areas of the country in recent decades. This is largely due to withdrawal for agricultural, industrial and urban use, in excess of annual recharge. In urban areas, apart from withdrawals for domestic and industrial use, housing and infrastructure such as roads prevent sufficient recharge.

In addition, some pollution of groundwater occurs due to leaching of stored hazardous waste and use of agricultural chemicals, in particular, pesticides. Contamination of groundwater is also due to geo-genic causes, such as leaching of arsenic and fluoride from natural deposits. Since groundwater is frequently a source of drinking water, its pollution and contamination leads to serious health impacts.

Main sources of water are:

- Surface water sources: Lakes impounding reservoirs, streams, seas, irrigation canals.

- Ground water sources: Springs, wells, infiltration wells.

Merits of ground water sources:

- The water quality is good and better than surface source.

- Being underground, the ground water supply has less chance of being contaminated by atmospheric pollution.

- The land above ground water source can be used for other purposes and has less environmental impacts.

- Prevention of water through evaporation is ensured and thus loss of water is reduced.

- Ground water supply is available and can even be maintained in deserted areas.

Demerits of ground water source:

- Modeling, analysis and calculation of ground water is less reliable and based on the past experience, thus posing high risk of uncertainty.

- Water obtained from ground water source is always pressure less. A pump is required to take the water out and is then again pumped for daily use.

- The transport/transmission of ground water is a problem and an expensive work. Water has to be surfaced or underground conduits are required.

- Boring and excavation for finding and using ground water is expensive work.

Impounding Reservoirs

It is a basin constructed in the valley of a stream or river for the purpose of holding stream flow so that the stored water may be used when supply is insufficient.

2.2.1 Floods

Floods have been a serious environmental hazard for centuries. However, the havoc raised by rivers overflowing their banks has become progressively more damaging, as people have deforested catchments and intensified use of river flood plains that once acted as safety valves.

Wetlands in the flood plains are nature's flood control systems into which the overfilled rivers could spill and act as a temporary sponge holding water and preventing fast flowing water from damaging the surrounding land. Deforestation in the Himalayas causes floods that year after year kill people, damage crops and destroy homes in the Ganges and its tributaries and the Brahmaputra.

Rivers can change their course during floods and tons of valuable soil is lost to the sea. As the forests are degraded, rainwater not a longer percolates that slowly into the subsoil but, runs off down in the mountainside bearing large amounts of topsoil.

This blocks the rivers temporarily but gives way as the pressure mounts allowing enormous quantities of water to wash down suddenly into the plains below. There, rivers swell, burst their banks and flood waters spread to engulf peoples farms and homes.

Cause and Effect of Floods

Whenever the magnitude of water flow exceeds the carrying capacity of channel within the banks, the excess of water overflows on the surroundings causes floods.

Causes of Floods

- Heavy rain, rainfall during cyclone causes floods.

- Sudden snow melt also raises the quantity of water in streams and causes flood.

- Sudden and excess release of compounded water behind dams.

- Human activities like construction of roads, building and parking space that covers the earth's surface prevents infiltration.

- Clearing of forest for agriculture has also increases severity of floods.

Effects of Floods

- Flood cause heavy suffering to people living in low lying areas because the houses and properties are washed away.

- Flood damage standing crops and livestock.

The unvarying and variable components of unvarying and variable characteristics:

	Stable Unvarying	Variable
Network Characteristics	Patters	Surface storage, under-drainage, channel length, contributing or source area etc.
Basin	Slope, attitude, shape, etc., of the basin	Arising from the interactions between climate, geology, soil type, vegetation cover, etc., which are manifested through the storage capacity of soil and bedrock, extent of infiltration and transmissibility of soil and bedrock, that are affected by anthropogenic activities.
Channel Characteristics	Slope, flood control and river regulation works	Roughness, load, shape, storage

Flood Management

- Encroachment of flood ways should be banned.

- Building walls prevent spilling out the flood water over flood plains.

- Diverting excess water through channels or canals to areas like lake, river, etc. where water is not sufficient.

- Build check dam on small streams, move building off the flood plains.

- Restore wetlands, replace ground course.

- River networking in the country also reduce flood.

- Flood forecast and flood warning are also given by the central water commission.

- Reduction of runoff by increasing infiltration through appropriate afforestation.

India Scenario-Floods

Next to Bangladesh, India is the most flood affected country in the world. Nearly 40 million hectares are affected by annual floods in India. 20% of which is present only in UR Next to UP or issa, Andhra Pradesh, Bihar.

2.2.2 Drought

The drought is due to lack or insufficiency of rain for an extended period that causes considerable hydrologic imbalances and consequently water shortages, stream flow reductions and depletion of groundwater levels and soil moisture. Drought is the most serious physical hazard to agriculture in nearly every part of the world.

Drought not only leads to serious economic consequences but also leaves behind untold human misery. Among all the natural disasters, drought affects largest number of people in the world. Shortage of water for even the basic needs is the main problem in the drought areas.

Even the shallow rooted crops do not grow in such areas. Getting sufficient drinking water is another problem needing immediate attention in these areas. Some of the measures like infiltration wells, underground dams, small watersheds, are being taken up to alleviate the sufferings of the people residing in the drought prone areas.

Certain advance techniques such as Cloud Seeding and Artificial Rains are also being tried with varying successes. However, these methods are quite expensive and unpredictable in their success. Scant rains for extensive periods also lead to ecological changes. Ultimately, Government has found reasonable remedies in the form of development of small watersheds in such areas.

In case of most arid regions of the world, the rains are unpredictable. This leads to periods when there is a serious scarcity of water to drink, it use in farms or provide for an urban and industrial use. Drought prone areas are faced with irregular periods of famine.

Agriculturists have no income in these bad years and as they have no steady income, they have a constant fear of droughts. India has 'Drought Prone Areas Development Programs', which are used in such areas to buffer the effects of droughts. Under these schemes, people are given wages in bad years to build the roads, minor irrigation works and plantation programs.

It is an unpredictable climatic condition and occurs due to the failure of one or more monsoons. It varies in frequency in different parts of our country. While it is not feasible to prevent the failure of the monsoon, good environmental management can reduce the ill effects.

The scarcity of water during the drought affects homes, agriculture and industry. It also leads to the food shortages and malnutrition that especially affects children. Several measures can be taken to minimize the serious impacts of a drought.

However this should be done as a preventive measure so that if the monsoons fail, its impact on the lives of the local people is minimized. In years when the monsoon is adequate, we use up the good supply of water without trying to conserve it and use the water judiciously.

Thus during a year when the rains are poor, there is no water even for drinking in the drought area. One of the factors that worsen the effect of drought is deforestation.

Once the hill slopes are denuded of forest cover rainwater rushes down the rivers and is lost. Forest cover permits the water to be held in the area allowing it to seep into the ground. This charges the underground stores of water in natural aquifers.

This can be used in drought years if the stores have been filled during a good monsoon. If water from an underground stores is overused the water table drops and vegetation suffers. This soil and water management and afforestation are long term measures which reduces the impact of drought.

2.2.3 Conflicts Over Water

Water conflict is a term describing a conflict between countries, states or groups over an access to water resources. The United Nations recognizes that water disputes result from opposing interests of water users, public or private.

A wide range of water conflicts appear throughout the history, though traditional wars waged over water alone. Instead, the water has historically been a source of tension and a factor in conflicts that start for other reasons.

Although, water conflicts arise for several reasons, including territorial disputes, a fight for resources and strategic advantage. A comprehensive online database of water related conflicts the Water Conflict Chronology has been developed by Pacific Institute. This database lists violence over water which are going back nearly 5,000 years.

These conflicts occur over both the freshwater and saltwater and both between and within nations. However, conflicts occur mostly over freshwater, because the freshwater resources are necessary, yet limited, they are the center of water disputes arising out of need for potable water and irrigation. As freshwater is a vital unevenly distributed natural resource, its availability often impacts the living and economic conditions of a country or region.

The lack of cost-effective water supply options in areas like Middle East, among other elements of water crises can put severe pressures on all water users, whether the corporate, government or individual leading to tension and possible aggression. Recent humanitarian catastrophes, such as the Rwandan Genocide or the war in Sudanese Darfur, have been linked back to water conflicts.

A recent report "Water Cooperation for a Secure World" published by Strategic Foresight Group concludes that active water cooperation between countries reduces the risk of war. This conclusion is obtained after examining the trans-boundary water relations in over 200 shared river basins in 148 countries, a growing number of water conflicts are sub-national.

Sustainable Water Management

'Save water' campaigns are essential to make people everywhere aware of dangers of water scarcity. A number of measures are to be taken for the better management of the world's water resources. These include measures such as:

- Building several small reservoirs instead of few mega projects.

- Soil management, micro catchment development and afforestation permits recharging of underground aquifers thus reducing the need for large dams.

- Preventing loss in Municipal pipes.

- Treating and recycling municipal waste water for an agricultural use.

- Preventing leakages from dams and canals.

- Develop small catchment dams and protect wetlands.

- Effective rain water harvesting in urban environments

- Pricing water at its real value makes people use it more responsibly and efficiently and reduces water wasting.

- Water conservation measures in agriculture namely using drip irrigation.

- In the deforested areas where land has been degraded, the soil management by bunding along the hill slopes and making 'nala' plugs can help to retain moisture and make it possible to revegetate degraded areas. Managing a river system is best done by leaving its course as undisturbed as possible. The dams and canals lead to major floods in the monsoon and the drainage of wetlands seriously affects areas that get flooded when there is high rainfall.

2.2.4 Dams: Benefits and Problems

The dams are the major structures in any reservoir from the point of view of structural importance, design details and cost. The dams are of different types depending on different criteria.

Depending on the material used for construction, dams can be: Earthen dams, rock fill dams, masonry dams, concrete dams, steel dams and timber dams.

It based on the design, the dams can be: Gravity dams, arch dams, buttress dams and multiple arch dams.

Similarly, based on the purpose, the dams are known as overflow dams and non-overflow dams.

The masonry and concrete dams are more or less leak proof and hence seepage is not possible. Although in the case of earthen and rock fill dams, seepage of water is expected, in good quantities and therefore, possibilities of water logging on downstream side will be the adverse effect on the environment.

The water resources projects are constructed to many purposes depending on the needs of people of the area to be served. Whenever the projects are developed to supply water for various purposes, the projects are termed as multipurpose projects.

The different purposes can be: irrigation and agriculture, hydropower generation, flood control, navigation, drinking water supply, water for Industries, recreation and amusement parks and afforestation.

All of the above purposes, irrigation and agriculture occupies higher priority, as the production of necessary food grains for one billion population of country is the primary concern to us.

To develop industries and other power needs, the next priority is the Hydropower development. The emphasis is increasing on the hydropower, as the natural resources for other forms of energy such as thermal are becoming scarce. Due to the rapid development of urban areas, scarcity of drinking water has surfaced in most of the cities. Hence, the present emphasis is on bringing water to the cities from storage reservoirs.

Thus slowly, water needs for drinking purposes is occupying the priority when compared to the other needs. In the flood plains, the problem of inundation of adjacent habituated areas is a priority, as during every flood the losses in respect of human and cattle life, crops, property and fertile soils, are bringing misery to people.

With every increasing movement of men and material, the transportation by navigation is also recognized as viable mode. In the recent times, water has been used for recreational purposes also. In some countries, water sports are gaining popularity. In order to develop greenery in many dry areas, Government is encouraging people to go in for tree plantations on a large scale resulting in afforestation.

Today there are more than 45,000 large dams around the world, which provide an important role in communities and economies that harness these water resources for their economic development. Current estimates suggest some 30-40% of irrigated land worldwide relies on dams.

Hydropower, another contender for the use of stored water, currently supplies 19% of the world's total electric power supply and is used in over 150 countries. The world's two most populous countries China and India have built around 57% of the world's large dams.

Dams Problems

- Serious impacts on riverine ecosystems.

- Fragmentation and physical transformation of rivers.

- Dislodging animal populations, damaging their habitat and cutting off their migration routes.

- Water logging and salinization of surrounding lands.

- Fishing and travel by boat disrupted.

- Social consequences of large dams due to its displacement of people.

- The emission of greenhouse gases from reservoirs due to rotting vegetation and carbon inflows from the catchment is a recently identified impact.

Large dams have had serious impacts on the livelihoods, lives, cultures and spiritual existence of indigenous and tribal peoples. They have suffered disproportionately from the negative impacts of the dams and often been excluded from sharing the benefits.

In India, of 16 to 18 million people displaced by dams, 40 to 50% were tribal people, who account for only 8% of our nation's one billion people. Conflicts over dams have heightened in the last two decades because of their social and environmental impacts and failure to achieve the targets for sticking to their costs as well as achieving promised benefits.

Recent examples show how failure to provide a transparent process that includes effective participation of the local people has prevented affected people from playing an active role in debating the pros and cons of the project and its alternatives. The loss of traditional, local controls over equitable distribution remains a major source of conflict.

2.3   Mineral Resources

Mineral resources are limited and thus, non-renewable. The materials extracted from soil has not been used for generating energy. The high consumption of minerals will lead to their depletion.

'Reserve Life Index' expresses the number of years of production at current annual rates for the pattern of production and consumption of earth's metals is not equitable in most of the cases. In 1991-92 USA, China, Japan and Russian Federation analysis has shown that India was the 6th largest producer of iron (34, 136, 000, metric tons).

Australia was the largest producer of aluminum and USA was the largest consumer of aluminum. Due to the high consumption of mineral resources, there is a huge environmental burden on the countries which extract the minerals. Extraction process, disposal of wastes of minerals have caused several environmental problems.

The land degradation by creating quarries, vast open pits and large amount of solid wastes have polluted water at several places. e.g:

- Iron minerals — Fe_2O_3 hematite, Fe_3O_4 magnetite, FeS_2.
- Copper minerals — Cu_2O—Cuprite, Cu_2S chalcosite, etc.
- Manganese minerals — MnO_2 pyrolusite, $MnSiO_3$ Rhodonite.
- Lead minerals — PbS Galena, $PbCO3$, Cerussite, Anglo-site.
- Zink minerals — ZnO Zincite, ZnS Blende etc.

The other groups of minerals are garnet, silicate, chlorite, mica and fuels.

Formation of Mineral Deposits

Concentration of the mineral at a particular spot, which can be extracted profitably and gives rise to a mineral deposit. The formation of these deposits is a very slow biological process; it even takes millions of years to develop as a mineral deposit.

Various Biological Processes

- Formation of mineral deposits is due to the biological decomposition of dead animals and organic matters.

- Mineral deposits are also formed due to evaporation of sea water.

- Mineral deposits are formed due to oxidation reduction reaction inside the earth.

Classification of Mineral Resources

U.S. Geological Survey divides non-renewable mineral resources into 3 categories:

- Identified resources.

- Undiscovered resources.

- Reserves.

1. Identified Resources

The location, existence, quantity and quality of these mineral resources are known by the direct geological evidence and measurements.

2. Undiscovered Resources

These mineral resources are assumed to exist on the basis of geological knowledge and they but their locations, quantity and quality are unknown.

3. Reserves

These mineral resources are identified resources, from which a usable minerals can be extracted profitably.

Uses of Minerals

- Development of industrial plants and machinery.

- Construction, housing, settlements.

- Generation of energy.

- Communication purposes.

- Medicinal purposes, particularly in Ayurveda system.

Classification of Minerals

Minerals are classified into two ways based on this composition and usage.

1. Based on Composition

Based on composition, minerals can be classified into two types:

i. Metallic Minerals:

Metallic minerals are the one from which various kinds of metals can be extracted.

Example: Iron, copper, zinc, etc.

ii. Non-Metallic Minerals:

Non-metallic minerals are the one from which various non-metallic compound can be extracted.

Example: Quartz, folds par, calcite.

2. Based on Usage

Based on usage, minerals are classified into two types:

i. Critical Minerals:

These are essential for the economic power of a country.

Example: Iron, Al, Cu and Au.

ii. Strategic Minerals:

These are required for the defense of a country.

Example: Manganese, cobalt.

Management of Mineral Resources

- The efficient use and protection of mineral resource.
- Modernization of the mining industries.
- Search for new deposit.
- Reuse and Recycling of the metals.
- Environmental impacts can be minimized by adopting the eco-friendly mining technology.

Use and Exploitation

The use of minerals varies greatly between countries. The greatest use of minerals occurs in developed countries. Like other natural resources, mineral deposits are un-evenly distributed around the earth.

Some countries are rich in mineral deposits and other countries have no deposits. The use of the mineral depends on its properties. For example aluminum is light but strong and durable so it is used for aircraft, shipping and car industries.

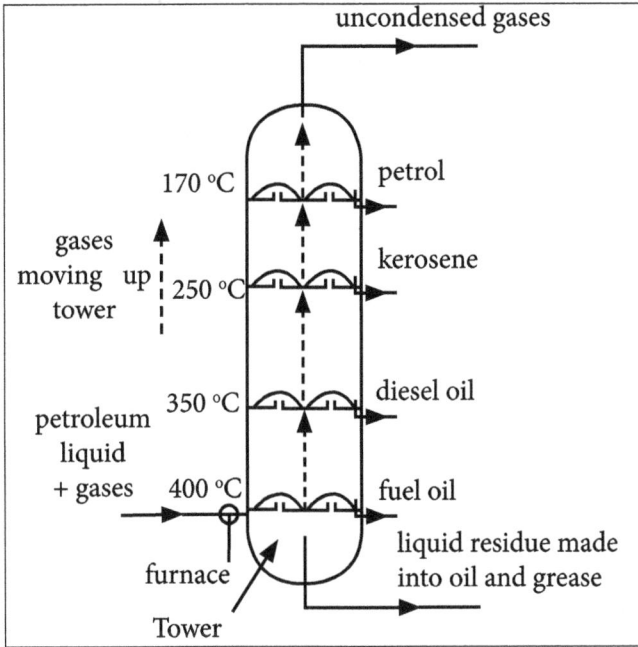

Fractional Distillation Tower.

Recovery of mineral resources has been with us for a long time. Early Paleolithic man found flint for arrowheads and clay for pottery before developing codes for warfare. And this was done without geologists for exploration, mining engineers for recovery or chemists for extraction techniques.

Tin and copper mines were necessary for a Bronze Age, gold, silver and gemstones adorned the wealthy of early civilizations and iron mining introduced a new age of man.

Human wealth basically comes from the agriculture, manufacturing and mineral re-sources. Our complex modern society is built around the exploitation and use of min-eral resources.

Since the future of humanity depends on mineral resources, we must understand that these resources have limits, we know supply of minerals will be used up early in the third millennium of our calendar.

Additionally, modern agriculture and the ability to feed an overpopulated world is de-pendent on mineral resources to construct the machines that tilt the soil, enrich it with mineral fertilizers and to transport the products.

We are now reaching limits of reserves for many minerals. Human population growth and increased modern industry are depleting our available resources at increasing rates. The pressure of human growth upon the planet's resources is a very real problem.

The consumption of natural resources proceeded at a phenomenal rate during the past hundred years and population and production increases cannot continue without increasing pollution and depletion of mineral resources.

The geometric rise of population as shown in the figure has been joined by a period of rapid industrialization, which has placed incredible pressure on natural resources. Limits of growth in the world are imposed not as much by pollution as by the depletion of natural resources.

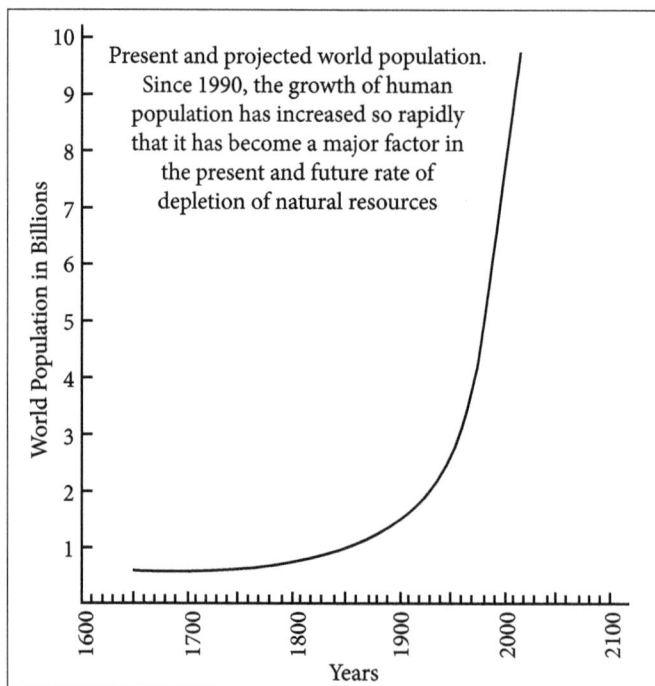

Present and projected world population. Since 1990, the growth of human population has increased so rapidly that it has become a major factor in the present and future rate of depletion of natural resources

Increase in World population.

As the industrialized nations of the world continue the rapid depletion of energy and mineral resources and resource-rich less-developed nations become increasingly aware of the value of their raw materials, resource driven conflicts will increase.

In the figure, we see that by about the middle of next century the critical factors come together to impose a drastic population reduction by catastrophe. We can avert this only if we embark on a planet-wide program of transition to a new physical, economic and social world that recognizes limits of growth of both population and resource use.

In a world that has finite mineral resources, exponential growth and expanding consumption is impossible. Fundamental adjustments must be made to the present growth culture to a steady-state system.

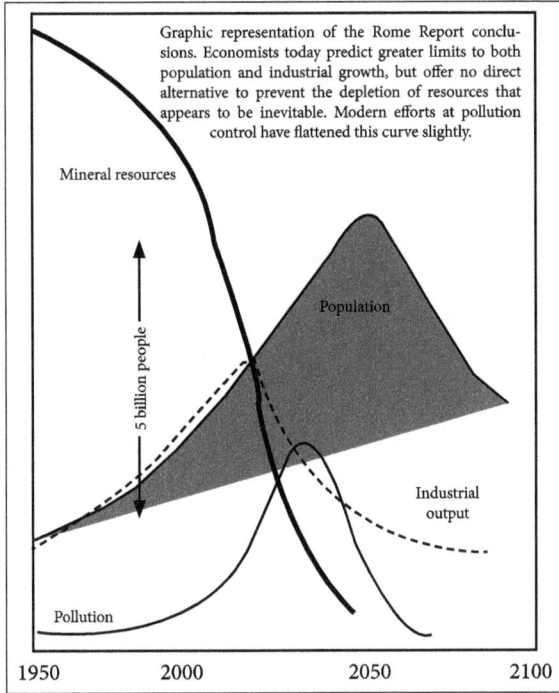

Graphic representation of the Rome Report conclusions. Economists today predict greater limits to both population and industrial growth, but offer no direct alternative to prevent the depletion of resources that appears to be inevitable. Modern efforts at pollution control have flattened this curve slightly.

Graphic representation of Rome report.

This will pose problems in that industrialized nations that are already feeling a loss in their standard of living and in non-industrialized nations that feel they have a right to achieve higher standards of living created by industrialization.

The population growth continues upward and the supply of resources continues to diminish. With the increasing shortages of many minerals, we have been driven to search for new sources.

2.3.1 Environmental Effects

Most important environmental concern arises from the impacts of extraction and processing of the minerals during mining, melting, roasting, etc.

Mining

Mining is the process of extraction of metals from a mineral deposit.

Types of Mining

The various types of mining are as follows:

i. Surface Mining

It is the process of extraction of raw materials from the near surface deposits.

ii. Underground Mining

The process of extraction of raw materials below the earth's surface is known as underground mining. It includes,

(a) Open-Pit Mining:

In this type of mining machines dig holes and remove the ores.

Example: Iron, Copper, Marble, etc.

(b) Dredging:

In dredging, chained buckets and drag lines are used, which scrap up the minerals from wider-water mineral deposit.

(c) Strip Mining:

In case of strip mining, the ore is stripped off by using bulldozers, stripping wheels.

Environmental Damage

The environmental damage, caused by mining activities are as follows:

1. De-vegetation and Defacing of Landscape

The top soil and the vegetation are removed from the mining area. Large scale deforestation or de-vegetation leads to a number of ecological losses and also landscape gets badly affected.

2. Ground Water Contamination

Mining disturbs and also pollutes the ground water. Usually sulphur, present as an impurity in many ores gets converted into disulphuric acid due to microbial action, which makes the water acidic. Some heavy metals also gets leached into ground water.

3. Surface Water Pollution

The drainage of acid from the mine often contaminates the nearby streams and lakes. The acidic water is harmful to many aquatic life. Radioactive substances like Uranium also contaminate the surface water and kill many aquatic animals.

4. Air Pollution

Melting and roasting are done to purify the metals, which emits enormous amounts of air pollutants damaging the nearby vegetation. The suspended particulate matter, arsenic particles, cadmium, etc., contaminate the atmosphere and public suffer from several health problems.

5. Subsidence of Land

It is mainly associated with underground mining. Subsidence of mining area results in cracks in houses, tilting of buildings, bending of railway tracks.

Environment Damages

- Mining activity not only destroys trees, it also pollutes soil, water and air with heavy metal toxins that are almost impossible to remove.
- Due to the continuous removal of minerals, forest covers, the trenches are formed on ground, leading to water logged area, which in turn contaminates the ground water.
- Destruction of natural habitat at the mine and waste disposal sites.
- During mining operations, the vibrations are developed, which leads to earth-quakes.
- Noise pollution is the another major problem from mining operations.
- When materials are disturbed in significant quantities during mining process, large quantities of sediments are transported by water erosion.
- Sometimes landslides may also occur as a result of continuous mining in forest area.
- Mining reduces the shape and size of the forest areas.

Environmental Impacts of Mining

The environmental responsibility of mining operations is the protection of the air, land and water. The mineral resources were developed in the United States for nearly two centuries with few environmental controls.

This is largely attributed to the fact that environmental impact was not understood or appreciated as it is today. In addition, the technology available during this period was not always able to prevent or control environmental damage.

Air

All methods of mining affect air quality. Particulate matter is released in surface mining when overburden is stripped from site and stored or returned to the pit. When the soil is removed, vegetation is also removed, exposing the soil to the weather, causing par-ticulates to become airborne through wind erosion and road traffic.

Particulate matter can be composed of noxious materials such as arsenic, cadmium and lead. In general, particulates affect human health adversely by contributing to illnesses relating to the respiratory tract, such as emphysema, but they also can be ingested or absorbed into the skin.

Land

Mining can cause physical disturbances to the landscape, creating eyesores such as waste rock piles and open pits. Such disturbances may contribute to the decline of wildlife and plant species in an area.

In addition, it is possible that many of the pre-mining surface features cannot be replaced after mining ceases. Mine subsidence can cause damage to buildings and roads.

Water

Water-pollution problems caused by mining include acid mine drainage, metal contamination and increased sediment levels in streams. Sources can include active or abandoned surface and underground mines, processing plants, waste-disposal areas, haulage roads or tailings ponds.

The sediments, typically from increased soil erosion, cause siltation or the smothering of streambeds. This siltation affects fisheries, swimming, domestic water supply, irrigation and other uses of streams.

Acid mine drainage (AMD) is a potentially severe pollution hazard that can contaminate surrounding soil, groundwater and surface water. The formation of acid mine drainage is a function of the geology, hydrology and mining technology employed at a mine site. The primary sources for the acid generation are sulfide minerals, such as pyrite (iron sulfide), which decompose in air and water.

Many of these sulfide minerals originate from waste rock removed from the mine or from tailings. If water infiltrates pyrite-laden rock in the presence of air, it can become acidified, often at a pH level of two or three.

This increased acidity in the water can destroy living organisms and corrode culverts, piers, boat hulls, pumps and other metal equipment in contact with the acid waters and render the water unacceptable for drinking or recreational use.

2.4   Food Resources

Food is an essential requirement for human survival. Each person has minimum food requirement. The main components of food are carbohydrates, fats, proteins, minerals and vitamins.

Types of Food Supply

Historically humans have a dependency on three systems for their food supply:

- Croplands: It mostly produces the grains and provides around 76% of the world's food.

  Example: Rice, Wheat, Maize.

- Range lands: It produces food mainly from grazing livestock and provide around 17% of the world's food.

  Example: Meat, Milk, Fruits.

- Oceans: Oceanic fisheries supply about 7% of the world's food.

  Example: Fish, Prawn, Crab, etc.

Food Problems

We know that 79% of the total area of the earth is covered with water. Only 21% of the earth surface is land, of which most of the areas are desert, forest, mountains, and barren areas. Only a minimal percentage of land is for cultivation, which in turn often does not suffice for the ever growing world population, as day by day the world population increases, while the cultivable land area decreases.

The environmental degradation like soil erosion, water logging, water pollution, salinity, affects agricultural lands. Urbanization is another problem in developing countries, which deteriorates the agricultural lands.

Since the food grains like wheat, rice, corn and vegetable like potato are the major food for the people all over the world, the food problem raises.

A key problem is the human activities which degrade most of the earth's net primary productivity which supports all life.

Urbanization is another problem in developing countries, which deteriorates the agricultural lands.

Under Nutrition and Malnutrition

- Nutrition (or) Nutritious (or) Nourished: To maintain good health and resist disease, we need large amount of the macro nutrients such as carbohydrates, proteins and fats and smaller amount of macro nutrients such as Vitamin A, C and E and minerals such as calcium, iron and iodine.

- The food and agriculture organization (FAO) of United Nation estimated that on an average, the minimum caloric intake on a global scale is 2,500 calories/day.

- Under Nutrition (or) Under Nourished: People who cannot afford enough food to meet the basic energy needs suffer from under nutrition. They receive less than 90% of the minimum dietary calories.

- Effect of under nutrition: Suffer from mental retardation and infections, diseases such as diarrhea.

- Malnutrition (or) Malnourished: Besides the minimum caloric intake we also need proteins, minerals, vitamins, iron and iodide. Deficiencies of lack of nutrition often lead to malnutrition resulting in several diseases.

Effect of Malnutrition

S. No.	Deficiency of Nutrient	Effects
1.	Proteins	Growth
2.	Iron	Anemia
3.	Iodine	Goiter, Cretinism
4.	Vitamin A	Blindness

Thus, chronically under nourished and malnourished people are diseases prone and are too weak to work or think clearly.

World Food Problems

- Environmental degradation like soil erosion, water pollution, water logging, salinity, affect agricultural lands.

- Urbanization is another problem in developing countries, which deteriorates agricultural lands.

- A key problem is the human activity, which degrade most of the world's net primary productivity which supports all life.

Changes Caused by Non-agriculture Activities

An engagement in nonagricultural activities in rural areas can be classified into survival-led or opportunity-led. Survival-led diversification would decrease inequality by increasing the incomes of poorer households and thus reduce poverty.

An opportunity-led diversification would increase inequality and has a minor effect on poverty, as it tends to be confined to nonpoor households. Using data from Western Kenya, we confirm the existence of the differently motivated diversification strategies. Yet, the poverty and inequality implications differ somewhat from the expectations.

Our findings indicate that in addition to asset constraints, rural households also face limited or relatively risky high return opportunities outside the agriculture.

2.4.1 Changes Caused by Modern Agriculture

Effects of Modern Agriculture involves the use of hybrid seeds, high-tech equipments, fertilizers, pesticides and irrigation water. Modern agriculture has succeeded significantly in increasing crop yield (green revolution).

Green Revolution: The introduction of scientifically bred or selected varieties of grain (rice, wheat, maize with high enough inputs of fertilizers and water can greatly increase crop yield).

Green revolution was possible because of:

- Use of high yielding varieties: In modern agriculture the use of high yielding varieties encourages monoculture. If a disease attacks, there is total devastation of the crop due to monoculture practice.

- Fertilizer and pesticides related problems: Fertilizers used in modern agriculture include phosphorus, potassium, nitrogen etc., Farmers use these fertilizers to increase the growth of crop. Excessive use of fertilizers may cause micronutrient imbalance.

Following are the problems related to the use of fertilizers:

- Pollution: Nitrate used in the fields gets leached down into deep soils and contaminate the ground water. The excessive levels of nitrate ions makes drinking water toxic, especially for infants. The disease is known as baby syndrome or metahemoglobinemia. This disease causes death of infants.

  ◦ Water pollution: Due to excessive use of NPK, run off from the fields to nearby water bodies causes nourishment of the lakes. Due to eutrophication, lakes get invaded by algal blooms. This causes death and decay of flora and fauna in lakes.

  ◦ Reduced oxygen content of soil: Excessive use of NPK fertilizer also reduces oxygen content of soil and alters its porosity.

  ◦ Reduced natural nitrogen production efficiency: Excessive use of NPK fertilizers causes the soil to become compact and less suitable for crop growth and reduces its natural ability to produce nitrogen in forms usable by plants.

- ○ Reduced humus production: Excessive use of organic and inorganic fertilizers also reduces humus production in the soil.

- Problems related to Pesticides: Various chemicals used to control pest population are called pesticides. Paul Mueller (1939) discovered insecticidal properties of DDT. There are thousands of pesticides used in agriculture.

The first generation pesticides include chemicals like arsenic, lead, mercury and sulphur to kill the pests while second generation pesticides are DDT. Although pesticides are protecting our crop yet they have created a number of problems which are discussed as:

- ○ Threat to wildlife: Wildlife gets destroyed due to the use of pesticides.

- ○ Development of genetic resistance: Some individuals of the pest species 'malty survive even after the pesticide spray and give rise to development of genetic resistant pest species.

- ○ Production of new pest: About twenty new species of pests are known which became resistant to all types of pesticides and are called as "Super-pest".

- ○ Death to non-target species: Many insecticides have broad spectrum poisons which not only kill the target species but also the non-target useful species.

- ○ Threats to human health: Due to excessive use of pesticides in contaminated food, human health is threatened.

2.4.2 Fertilizer-pesticide Problems

Problems of Pesticides on Modern Agriculture

In order to improve the crop yield, lots of pesticides are used in the agriculture.

1. First generation pesticides: Sulphur, arsenic, lead or mercury are used to kill the pests.

2. Second generation pesticides: DDT is used to kill the pests. Although these pesticides protect our crops from huge losses due to pests, they produce number of side effects such as:

- Death of non-target organisms.

- Producing new pests.

- Bio magnification.

- Risk of cancer.

Effect of Pesticides

- It directly acts as Carcinogens.

- It indirectly suppress the immune system, the soil and contaminate the ground water. The nitrate concentration in the water gets increased. When the nitrate concentration exceeds 25 mg/liter, they cause serious health problem called, "Blue Baby Syndrome". This disease affects infants and leads even to death.

Derived Qualities of an Ideal Pesticide

- It must be a biodegradable.

- It should not produce new pests.

- It should not produce any toxic pesticide vapor.

- Excessive synthetic pesticide should not be used.

- Chlorinated pesticides and phosphate pesticides are hazardous, so they should not be used.

2.4.3 Water Logging

Water logging is the saturation of soil with irrigation water so that the water table rises close to the surface. Worldwide, about one-tenth of all irrigated land suffers from water logging. Under waterlogging conditions, porous spaces in the soil are drenched with water and soil air gets depleted.

The water table rises and roots of plants do not get adequate air for respiration. Crop plants get lodged, mechanical strength of the soil declines and results in low crop yield. Water logging can be controlled by preventing excessive irrigation, sub-surface drainage technology and bio-drainage with trees like Eucalyptus.

Problems in Water Logging

During water logged conditions, the soil gets filled with water and the soil air gets depleted. In such conditions the roots of the plants do not get enough air for respiration. So, mechanical strength of the soil decreases and crop yield fails.

Causes of Water Logging

- Heavy rain.

- Excessive water supply to the crop lands.

- Poor drainage.

Remedy

Preventing excessive irrigation, sub surface drainage technology and bio drainage by trees like eucalyptus tree are some method of preventing water logging.

2.4.4 Salinity

Salinity is the quantity of salt dissolved in a given volume of water. Accumulation of salts in soil can eventually make the soil incapable of supporting plant growth. This is called salinization. One-third of the world's irrigated land is now affected by salinity.

In India, about seven million hectares of land is affected by salinity. Saline soils are characterized by having electrical conductivity more than 4 ds/m and pH exceeds 8.0 and the exchangeable sodium percentage is more than 15 percent.

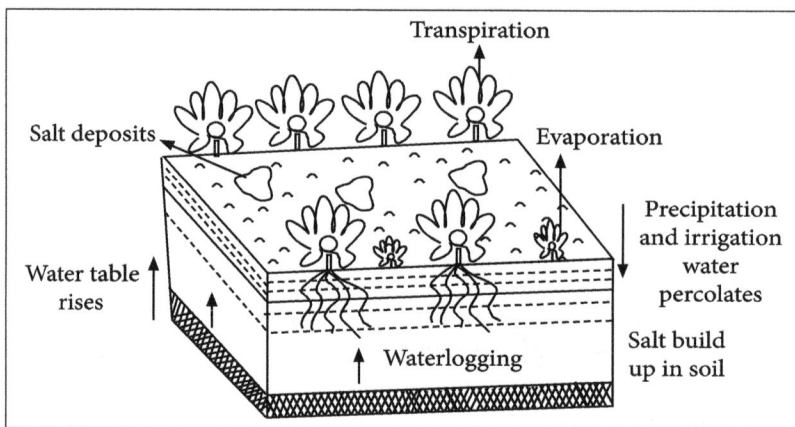

Waterlogging and salinization.

Major cause of salinization is excessive irrigation. About 20 percent of the world's crop-land receives canal and ground water which is usually contaminated by dissolved salts. During dry climate, water evaporates leaving behind salts in the top soil profile.

Salinization Cause

- Decreased crop yield.

- Stunted crop growth and at times kills them.

- Land becomes unproductive.

- Hazardous to wild life, especially water fowl.

Remedy

- The salt deposit is removed by flushing them out by applying higher grade quality water to such soils.

- Using sub surface drainage system the salt water is flushed out slowly.

Case Studies

According to in (1985) pesticides related deaths in developing countries are estimated at 10,000 per year and about 1.5-2 million people suffer from acute pesticides poisoning.

Some of the most toxic biocides are DDT (dichloro-diphenyltrichloroethane), BHC (benzene he-xa chloride), chlordane, heptachlor, methoxychlor, toxaphene, aldrin, endrin) and PCBs (Polychlorinated biphenyls).

Measurable amount of DDT residues are found in air, soil water and at several thousand of kms from the point where it entered the bio-system/ecosystem. DDT entering ponds/lakes is taken by plants, then reaches zooplanktons then to minnows feeding a 300 planktons, then to fish and finally to our body and the body of the birds. DDT concentration continuously increases in successive trophic level in a food chain.

This phenomenon is called Biological Magnification or Biological Amplification. Besides DDT there are certain heavy metals like lead, mercury, copper, also show similar behavior.

Similarly the radio nucleosides as strontium 90 follow biological magnification. In an island in USA, after regular DDT spray for many years, the population of fish eating birds began to decline. Several children died in Rajasthan due to nitrate poisoning.

In 1976, there was a case of nitrate poisoning of cattle in Nagpur. Children born today have to start life with a body burden of pesticides which increases with age. There is evidence that such chronic accumulation of pesticides played a role in Kidney malfunctioning. Excess of amino acids in blood and urine causes electroencephalogram, abnormalities of brain tissue, blood abnormalities, etc.

The minamata epidemic caused several deaths in Japan and Sweden. The tragedy occurred due to consumption of heavily mercury contaminated fish (27 to 102 ppm).

2.5   Energy Resources

Energy has been defined as the capacity to do work. Energy exists in two forms: potential and kinetic energy. Chemical energy is the source of energy required by all living organisms, which they acquire through their food, by the radiant energy of the sun.

The electromagnetic waves from the sun are absorbed by the plants; it is then converted to biochemical energy stored in food of living organisms. The energy flow in any ecosystem should be unidirectional. This energy flow is based on two important laws of Thermodynamics namely the first law of Thermodynamics and the second law of Thermodynamics.

The main source of energy is the sun and 57% of the sun's energy is absorbed in the atmosphere and scattered in space, 36% is expend to evaporated water of the 88% of sun light striking a plant surface, 10-16% is reflected, 5% is transmitted and 80-85% is absorbed. On an average only 2% of light energy is utilized for photosynthesis by plants, the rest is all transformed into heat energy.

Autotrophs of an ecosystem can fix energy from inorganic sources to organic molecules. Heterostrophic cannot obtain energy from abiotic sources and thus depend upon autotrophs. Consumers obtain energy from living organisms, while decomposers obtain it from dead organisms.

Energy in fact is the most important input for development. It aims at human welfare covering household, agriculture, transport and industrial complexes. Just like all the other resources, energy resources are also renewable and non-renewable.

Growing Energy Needs

Energy is an important aspect of development. As the civilization proceeds, energy needs increases. Our day to day chores require abundance of commercial energy.

More than 80% of total energy produced is consumed by developed countries which are occupied by only 30% of the total human population on the earth. Energy is needed for transportation, industries lighting etc. Energy is required at every walk of our life.

90% of the world's production of commercial energy is met by the conventional energy resources like petroleum oil, Natural gas coal etc. while only 10% is met by Hydro-electric and nuclear power. With the growing population each day, the energy requirements are also rising steeply.

As the conventional sources are not enough to meet the growing demand for energy and there is a risk of their exhaustion hence a number of alternatives are being used and thus the alternative energy resources also play vital role to provide us energy now and in near future. Ironically, only 20% of the energy is consumed by the 70% of the world population that lives in developing and social countries.

Fossil Fuels

Organic fuel found or derived under the earth's surface is a chief source of fuel. Fossil fuel deposits do not form overnight, it takes millions of years to generate a fossil fuel reserve, thus they may be classified under the category of non-renewable resources, coal and petroleum oil/minerals oil under fossil fuel. The chief sources of energy in many countries are fossil fuel, but the fossil fuel deposits are fast depleting and may one day exhaust.

1. Coal

About 600 billion tons of coal lies under the earth and by now more than 200 billion tons has been used .Coal a prime source of industrial energy is also a raw material. 60% of our countries commercial energy requirements are met by coal. Bihar or issa, West Bengal, U.P., A.P. and Maharashtra are the major coal producing states of India. Per capita consumption of coal has increased from 135 kg to 225 kg.

2. Oil and Natural Gas

Sedimentary rocks containing plants and animals remains about 10 to 20 crore year old are the source of minerals oil.

Petroleum oil is also very unevenly distributed over space. Only six regions in the world are found to be rich in mineral oils. USA, Mexico, USSR and West Asian regions, Oil production has also gone down with time.

India has large proportion of tertiary rocks and alluvial deposits particularly in ex-tra peninsular India. It cover northern plains in the Ganga-Brahamaputra Valley, west strips, together with their off shore continental shelf. (Bombay High), the plains of Gujarat, the Thar desert and the area around Andaman and Nicobar island. (Earlier it was Assam where oil wells were dug)."

Natural gas comes hand in hand with minerals oil. It can both be used as a sources of energy and also an industrial raw material. It can be well used in fertilizer plants. There are 12 refineries in India.

The liquefied petroleum gas (LPG), also called the cooking gas is now a very common domestic fuel.

3. Hydropower

Water energy is the most conventional renew-able energy source. Energy is obtained from water flow or falling water from a height. It is the main reason why in 10th and 19th century most of the industries were set near water falls. Hydropower converts power of the falling waters to electrical power which can be transmitted to long dis-tance through wires and cables.

This form of energy however cannot be stored for future use, thus, markets are to be fixed before generation of this energy. Apart from India, Switzerland, Canada, Sweden and New Zea-land, harness their water resources from this form of energy.

4. Nuclear Power

A small amount of radioactive material can be a source of enormous amount of energy. This, thus becomes a principal resources when fossil fuels are depleting at such a high rate. For instance, one ten of U235 would provide energy equal to that generated by 3 million ton of coal or 12 million barrels of oil. For the generation of atomic/nuclear energy "nuclear reactors" are required.

The nuclear reactors must be used with utmost caution. India also has rich reserves of nuclear fuels in Bihar, Rajasthan, Kerala, Tamil Nadu, Karnataka etc. The implications of such energy sources are severe and due care has to be exercised while disposing of the waste coming from such plants.

Renewable and Non-renewable Energy Sources

Renewable Energy Resources

The resources have unlimited stock in the nature. These are mostly biomass based resources. Once consumed, it can be renewed relatively at a short-period of time. They are inexhaustible. They include fuel wood, petro plants, plant biomass, animals' dung, solar, wind, water, geothermal and dendohedron energy. These can reproduce themselves and can be harvested continuously through a sustained proper planning and management.

Non-Renewable/Exhaustible Energy Resources

They are limited in stock in nature and get regenerated over a long period of time. They may be exhausted one day due to excessive use: Coal, mineral oil, natural gas and nuclear power are some example as given,

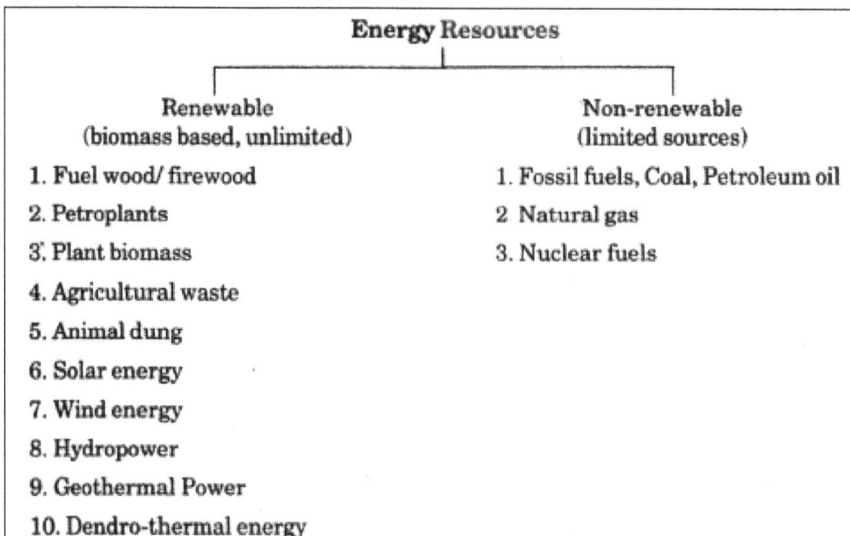

Energy Resources	
Renewable (biomass based, unlimited)	**Non-renewable** (limited sources)
1. Fuel wood/ firewood	1. Fossil fuels, Coal, Petroleum oil
2. Petroplants	2 Natural gas
3. Plant biomass	3. Nuclear fuels
4. Agricultural waste	
5. Animal dung	
6. Solar energy	
7. Wind energy	
8. Hydropower	
9. Geothermal Power	
10. Dendro-thermal energy	

Sources of energy may be classified as conventional and non- conventional energy resource.

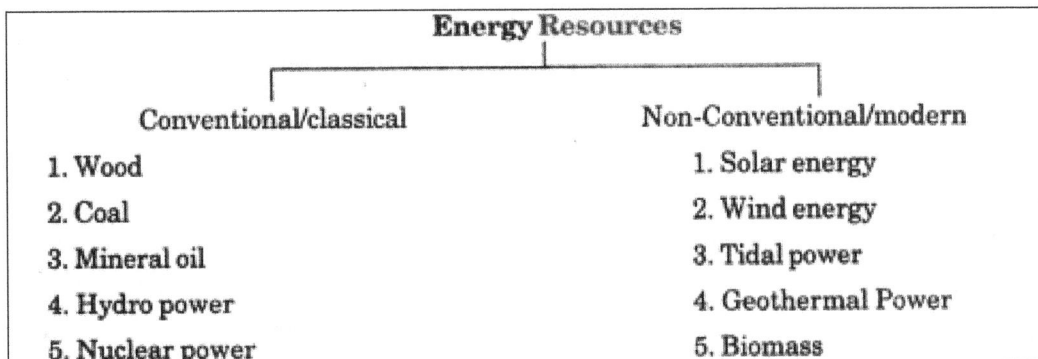

Energy Resources	
Conventional/classical	**Non-Conventional/modern**
1. Wood	1. Solar energy
2. Coal	2. Wind energy
3. Mineral oil	3. Tidal power
4. Hydro power	4. Geothermal Power
5. Nuclear power	5. Biomass

Use of Alternate Energy Sources

The world is plagued by severe energy crisis which derail the global economy and disrupt the environmental support system. It is also beyond the capacity of the global biosphere to absorb the emission of a fossil fuel based energy system. Concentration of Carbon dioxide in the atmosphere change temperature in global atmosphere-Global warming.

The reduced availability of fossil/conventional fuel especially petrochemical and the limited capacity of the world to cope with the overwhelming pollution caused by fossil fuels are the two major considerations that have forced the world to seek an alternative solution.

It is hoped that in times to come a clean, non-polluting sustainable energy system which will be very different from fossil fuel will ultimately come into existence.

Some of the alternative/non-conventional sources of energy are:

- Solar energy resources.

- Wind energy resources.

- Geothermal energy resources.

- Tidal energy resources.

- Biomass.

1. Solar Cells or Photovoltaic Cells or PV Cells

Solar cells consists of a p-type semiconductor and n-type semiconductor. These are in close contact with each other. When solar rays fall on the top layer of p-type semiconductor, the electrons from the valence band get promoted to the conduction band

and cross the p-n junctions into n-type semiconductor, thereby potential difference between two layers is created which causes flow of electrons.

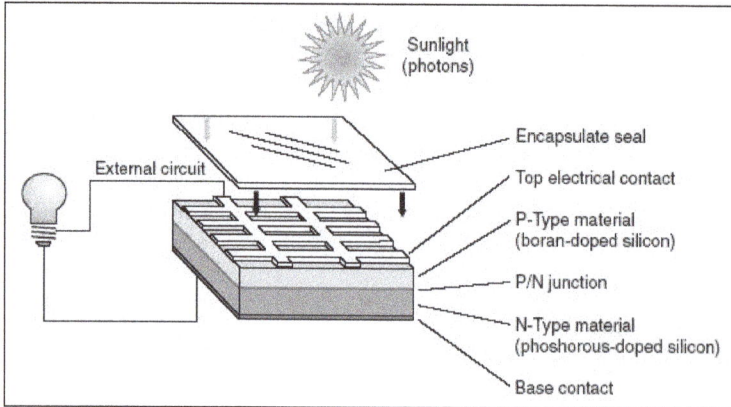

Solar cell.

Uses:

Solar cells are used in electronic watches, calculators, water pumps, street lights and to run radio and TVs.

Solar Battery

When a large number of solar cells are connected in series it form a solar battery. Solar battery produces electricity capable of running water pump, to run street-light etc. They are used in remote areas where conventional electricity supply is a problem.

2. Solar Water Heater

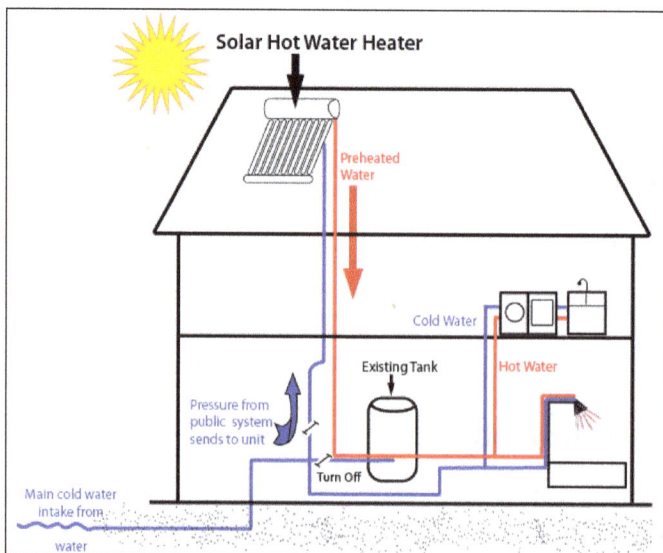

Solar water heater.

It consists of an insulated box inside of which is painted with black paint. It is also provided with a glass lid to receive and store the solar heat. Inside the box, it has a black painted copper coil, through which cold water is allowed to flow in, which gets heated up and flows out into a storage tank. From the storage tank water is then supplied through pipes.

Significance of Solar Energy

- Solar water heaters, cookers, require neither fuel nor attention while cooking the food.

- Solar cells are free from air and sound pollution.

- Solar cells can be used in remote and isolated areas, forests, hilly regions.

3. Solar Heat Collectors

Solar heat collectors consists of the natural materials like stones, bricks (or) materials like glass, which can absorb heat during the day time and release it slowly at night.

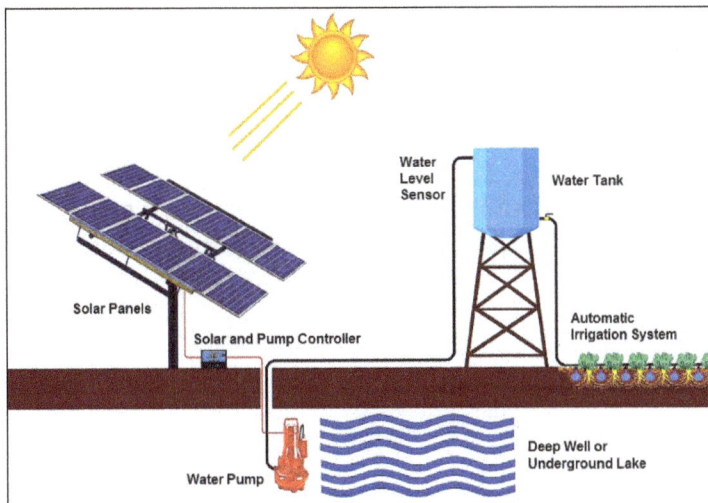

Solar pump run by solar cells.

Uses:

It is generally used in cold places, where the houses are kept in hot condition using solar heat collectors.

Wind Energy

Moving air is called wind. Energy recovered from the force of the wind is called wind energy. The energy possessed by wind is because of its high speed. The wind energy is harnessed by making use of wind mills.

1. Wind Mills

The strike on the blades of the wind will make it rotating continuously. The rotational motion of blades drives a number of machines like water pump, flour mills and electric generators.

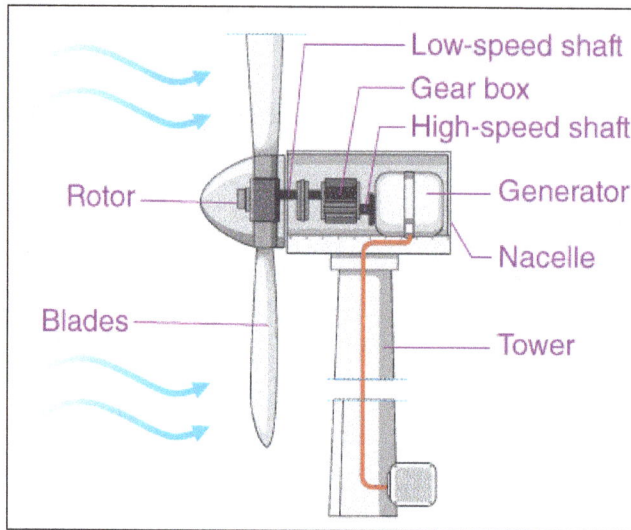

Wind mill.

Wind Forms

When a large number of wind mills are installed and joined together in definite pattern it forms a wind farm. The wind farms produce large amount of electricity.

Condition:

The minimum speed required for satisfactory working of a wind generator is 15 km/hr.

Advantages

- It is very cheap.

- It does not cause any air pollution.

Significance of Wind Energy

- It is made available easily in many off-shore, on-shore and remote areas.

- The generation period of wind energy is low and power generation starts from commissioning.

- It is recommended to broaden the nation's energy options for new energy sources.

Ocean Energy

Ocean can also be used for generating energy in the following ways.

Ocean Thermal Energy

There is often a large temperature difference between the surface level and deeper level of the tropical oceans. This temperature difference can be utilized to generate electricity. The energy available due to the difference in temperature of water is called thermal energy.

Condition:

The temperature difference of 20°C or more is required between surface water and ground water.

Process

The warm surface water of ocean is used to boil a low boiling liquid like ammonia. The high vapor pressure of the generates electricity. The cold water from the deeper ocean is pumped to cool and condense the vapor into liquid.

Significance of OTE

- OTE is continuous, renewable and pollution free.
- The use of cold deep water as the chiller fluid in air conditioning, has also been proposed.
- Electric power generated by OTE can be used to produce hydrogen.

Geothermal Energy

The temperature of the earth increases at a rate of 20 - 75°C per km, when we move down the earth surface. High temperature and high pressure steam fields exists below the earth's surface in many places. The energy harnessed from the high temperature present inside the earth is called geothermal energy.

1. Natural Geysers

In some of the places, hot water (or) steam comes out of the ground through cracks naturally in the form of natural geysers.

2. Artificial Geysers

In some of the places, we can artificially dig a hole up to the hot region and by sending a pipe in it, we can make hot water or steam to rush out through the pipe with very high pressure.

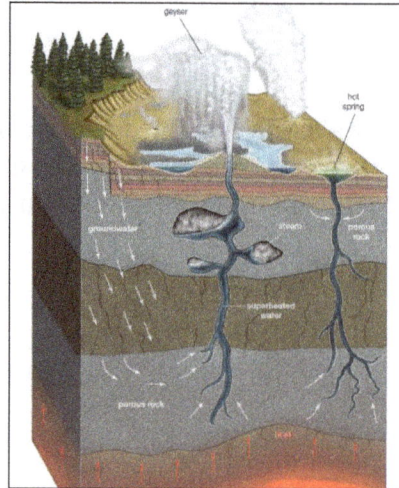

Artificial geysers.

Thus, the hot water (or) steam coming out from the natural (or) artificial geysers is allowed to rotate the turbine of a generator to produce electricity.

Significance of Geo-Thermal Energy

- The power generation level is higher for geothermal than for solar and wind energies.

- Geothermal power plants can be brought on line more quickly than most other energy sources.

- GTE is effectively and efficiently used for direct uses such as hot water bath, resorts, aquaculture green houses.

Tidal Energy or Tidal Power

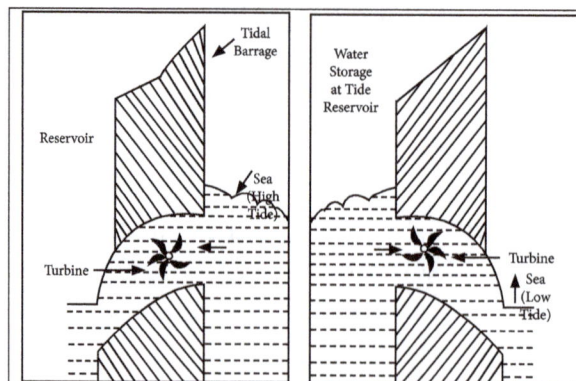

Water flows into the reservoir from sea (b) Water flows out from the reservoir to the sea.

Ocean tides, produced by gravitational forces of sun and moon, contain enormous amount of energy. The "high tide" and "low tide" Can refer the rise and fall of water in the oceans. The tidal energy shall be harnessed by constructing a tidal barrage.

During high tide, the seawater is allowed to flow into the barrage reservoir and rotates the turbine which in turn produces electricity by rotating the generators.

During low tide, when the sea level is low, the sea water stored in the barrage reservoir is allowed to flow in and in turn rotate the turbine again:

- Water flows into the reservoir from sea.

- Water flows out from the reservoir to the sea.

Significance of Tidal Energy

- Tidal power plants do not require large areas of valuable lands as they are on the bays or estuaries.

- As the sea water is inexhaustible, it is completely independent of the uncertainty of precipitations.

- It is pollution free energy source as it does not use any fuel and also does not produce any wastes.

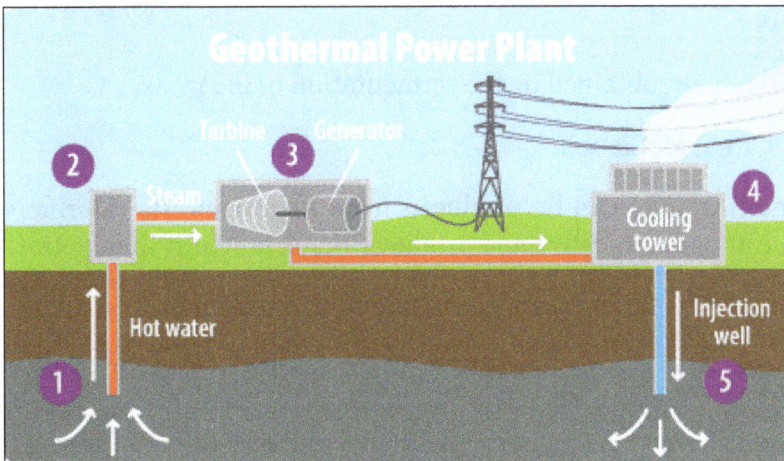

Geothermal power plants.

Biomass Energy

Biomass is the organic matter, produced by plants or animals used as sources of energy. Most of the biomass is burned directly for heating and cooling and industrial purpose.

Examples: Wood, crop residues, seeds, cattle dung, sewage, agricultural wastes, etc. Biomass energies are of any one of the following types.

1. Bio Gas

Bio gas is a mixture of gases such as methane, carbon dioxide, hydrogen sulfide, etc. It contains about 65% of methane gas as a major constituent.

Bio gas is obtained by the anaerobic fermentation of animal dung or plant wastes in the presence of water.

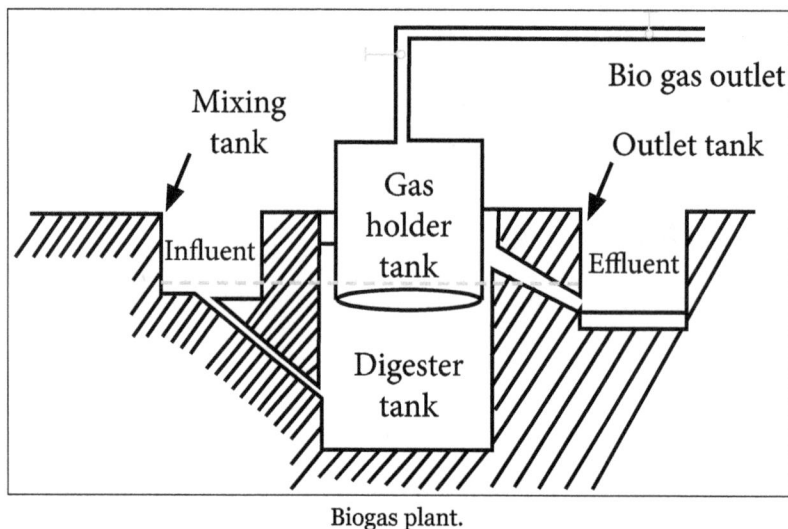

Biogas plant.

2. Bio Fuels

Bio fuels are the fuels, obtained by the fermentation of bio mass.

Examples: Ethanol, Methanol:

- Ethanol: Ethanol can be produced from sugarcane. Its calorific value is less when compared to petrol and produces much less heat than petrol.

- Methanol: Methanol can be obtained from ethanol or sugar containing plants. Its calorific value is also too low when compared to diesel and gasoline.

- Gasohol: Gasohol is a mixture of ethanol and gasoline. In India, a trial is being carried out to use Gasohol in cars and buses.

3. Hydrogen Fuel

Hydrogen can be produced by thermal dissociation or photolysis or electrolysis of water. It possess high calorific value. It is nonpolluting, because the combustion product is water.

$$2H_2 + O_2 \rightarrow 2H_2O + 150$$

Disadvantages of Hydrogen Fuel

- Safe handling is required.

- Hydrogen is difficult to store and transport.

- Hydrogen is highly inflammable and explosive in nature.

Conventional (or) Non-renewable Energy Resources

Coal: Coal is a solid fossil fuel formed in several stages as buried remains of land plants that lived 300–400 million years ago were subjected to intense heat and pressure over millions of years.

Various Stages of Coal

The various stages of coal during the coalification of wood is:

Wood → Peat → Lignite → Bituminous

The carbonate content of anthracitic is 90% and its calorific value is 8700 k.cal. The carbon content of bituminous, lignite, pert and wood are 80, 70, 60 and 50% respectively.

Disadvantages

- When coal is burnt it produces CO_2, which causes global warming.

- Since coal contains impurities like S and N, it produces toxic gases during burning.

1. Natural Gas

Natural gas is found above the oil in oil well. It is a mixture of 50-90% methane and a small amount of other hydrocarbons. Its calorific value ranges from 12,000- 14,000 k.cal/m³.

2. Dry Gas

If the natural gas contains lower hydrocarbons like methane and ethane, it is called dry gas.

3. Wet Gas

If the natural gas contains higher hydrocarbons like propane, butane along with methane it is called wet gas.

Like petroleum oil, natural gas can also be formed by the decomposition of dead animals and plant, that were buried under lake and ocean, at high temperature and pressure for million of years.

Crude Oil

It is a mixture of hydrocarbons that are formed from plants and animals that lived millions of years ago. It is a fossil fuel and it exists in liquid form in underground pools or reservoirs, in tiny spaces within sedimentary rocks and near the surface in tar sands. Petroleum products are fuels made from crude oil and other hydrocarbons contained in natural gas. Petroleum products can also be made from coal, natural gas and biomass.

Case Study

Solar energy is used to provide 65% of domestic hot water in Israel. The Lux international company, California extracts much of its power requirements using wind energy and solar through collector process.

Koyna dam (Maharastra, India) is considered to be the major cause of earth quake near it. Coastal India has been reported to have used wind power for power production.

The Hiroshima and Nagasaki nuclear hazards can never be forgotten which still remind one of the misuse of nuclear power.

2.6   Land Resources

The occupation is nearly 20% of the earth surface, measuring almost up to about 13000 million hectares of area. The houses, roads and factories occupy nearly one third of the land. The forests occupy another one third of the land.

The rest of land is used for and for meadows and pastures. The soil forms the surface layer of the land which covers more than the 80% of land. The soil is defined as a natural body which keeps on changing and even allows the plants to grow.

It consists of organic and inorganic materials. This definition is given by Buckman and Brady. The branch of science which can deal with the formation and distribution of soil in the different parts of the world is referred as a pedology. The professional which deals with the soil is known as the pedologist.

The inorganic component in the soil is 45 percent and the organic component in the soil is 5%. The water component in the soil is 25% and air component in the soil is 25%. The soil particles have fine spaces which are known as pore spaces. These are also known as the interstices.

They contain air and water along with the dissolved substances. The water and air content in the soil is inversely related to each other. The more is the water content lesser is the space for air to exist. The soil has both animals and plants. The micro flora consists of heterotrophic and autotrophic bacteria.

It also contains the fungi and algae. The heterotrophic bacterium consists of nitrogen and non-nitrogen fixing bacteria. Nitrogen fixing bacteria can be symbiotic, non-symbiotic, aerobic and anaerobic. Non nitrogen fixing bacteria can either be aerobic or anaerobic. The fungus includes yeast and mushrooms.

Algae can be red or brown or green. The fauna can be micro or macro. The micro fauna includes protozoa and nematodes. The macro fauna includes the mites, termites, earthworm, snails and mice. The soil has different types of soil particles. The mineral composition of the rock determines them along with the size of particles.

It includes gravel particles, sand, silt and clay particles. The gravel particles consist of mainly small stones and have a few sand particles and are used to make roads. The sand particles consist of pores and are aerated. They can hold little bit of water and are made up of large quartz. The silt particles are moved by the help of water. They are left at the bank of river. They are inert and are made up of large quartz.

The clay particles contains nutritive salts and have ability to retain the water. They are not inert and react chemically. Some of their pure forms are not suitable for growth of plants as they form a non-penetrable mass. The other components of soil mix with the clay particle and form a granular soil.

This type of soil is ideal for the cultivation. It contains pores as well as has the ability to hold water. It also contains nutritive salts.

The loamy soil is made up of clay, silt and sand. The proportion of clay is least and is half as compared to the silt and sand. The silt and the sand are twice and equal in the proportion. It is also a good soil for the growth of plants as it has pores as well as has the ability to hold water. It also contains some nutritive salts.

There are many factors which control the nature of soil. They are porosity, water holding capacity and texture. They come under the physical nature of soil. The chemical nature of the soil is governed by the salt content, inorganic and organic content includes certain metals.

The climate, topography and organisms also play a vital role in deciding the nature of soil. The half decayed and half synthesized part of organic material in the soil forms the humus. It contains nutrients and help in growth. It has the ability to absorb the heat and warm the soil. It makes the soil granular by its porosity and water holding capacity.

2.6.1 Land Degradation

Land degradation is the process of deterioration of soil or loss of fertility of the soil. It is the reduction in capacity of the land to provide ecosystem goods and services and assure its functions over a period of time.

Land degradation affects large areas and many people in the dryland regions. Increased population pressures and excessive human expansion into dry lands during long wet periods leave an increasing number of people stranded there during dry periods. The transfer of critical production elements to other uses through the introduction of irrigated and non-irrigated cash crops and the use of water for the industrial and urban purposes at the expense of rural agricultural producer break links in the traditional production chains in dry lands.

The removal of protective cover to reduce the competition for ploughing, water and nutrients, heavy grazing and deforestation leave the soil highly vulnerable to wind erosion particularly during severe droughts. Heavy grazing around water points or during long droughts prevents the regrowth of vegetation or favors only unpalatable shrubs.

Harmful Effects of Land (Soil) Degradation

- The soil texture and soil structure are deteriorated.

- Increase in water logging, salinity, alkalinity and acidity problems.

- Loss of soil fertility, due to loss of invaluable nutrients.

- Loss of economic social and biodiversity.

Causes of Land Degradation

1. Population

As population increases, more land is needed, for producing food, fiber and fuel wood. Hence there is more and more pressure for the limited land resources, which are getting degraded due to over exploitation.

2. Urbanization

The increased urbanization due to population growth reduces extent of agricultural land. To compensate the loss of agricultural land, new lands comprising natural ecosystems such as forests are cleared. Thus the urbanization leads to deforestation, which in turn affects millions of plant and animal species.

3. Fertilizers and Pesticides

Increased applications of fertilizers and pesticides are needed to increase farm output in the new lands, which again leads to pollution of land and water and soil degradation.

4. Damage of Top Soil

Increase in soil food production generally leads to damage of top soil through nutrient depletion.

5. Water Logging

Water Logging, soil erosion, desalination and contamination of the soil with Industrial wastes all cause land degradation.

2.6.2 Man Induced Landslides

The causes of man induced landslides are as follows:

- Excavation.

- Loading.

- Draw-down.

- Land use (e.g., construction of roads, houses etc.).

- Water management.

- Mining.

- Quarrying.

- Vibration.

- Water leakage.

- Deforestation.

- Land use pattern.

- Pollution.

Human Factors in Man Induced Landslides

These basically include human activities like construction of roads, buildings, dams, etc. These have strong bearing on the Man Induced Landslides. First road in Himalayan region was introduced way back in British colonial period (Singh and Ghai, 1996). It was, however, only after the Indo-China war of 1962 that the road construction was intensified in the Himalayas.

Thereafter, Indian engineers blasted gigantic networks of road and communication facilities deep into hills of the Himalaya. In NDBR, the history of road construction began in sixties. In 1964, for first time, roads were constructed in the reserve. Local people reported that initially about 80 km of roads were constructed in the region. About 50 km of road were further added to the existing network of roads in 2000.

Presently, road length is about 135 km. Now the road in Mana valley has been declared as national highway due to which slopes are now in process of being over-modified. Thus, sheer stress factors are exceeding sheer strength factors and landslides are increasing. After the introduction of roads, landslides have become very frequent in the reserve. The correlation between construction of roads and landslides is positive. It has the three stage relations.

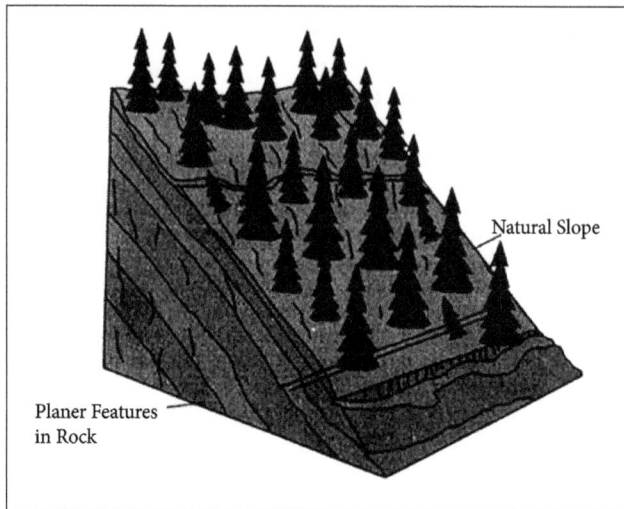

Stage I: Natural slopes - Premodified stage.

First stage, which is the premodified stage of natural slope, the steady-state condition is found. Then, the natural slopes are modified or undercut for the construction of roads and dams, etc. Thus, steady-state condition is disabled.

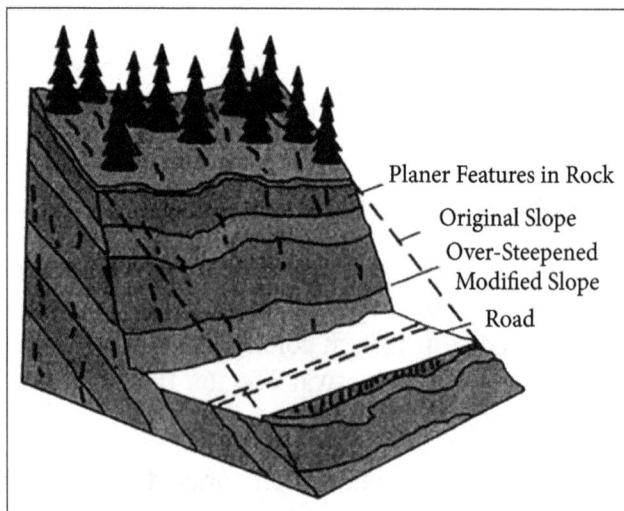

Stage II: Disabled steady-state condition.

Then landslides take place so that steady-state condition is reestablished. Therefore, landslides can be referred as nature's rule of equilibrium.

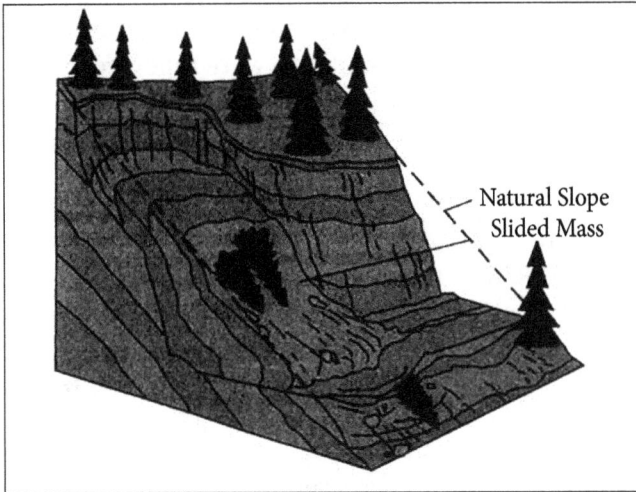

Stage III: Nature Re-establishing steady-state condition.

Built-up land has also increased in the recent past due to which natural slopes have been over-modified. It has also led to deforestation in the reserve. Thus, steady-state condition has disabled in many areas, which have eventually resulted in increased frequency of landslides.

Strong positive correlation between occurrence of landslides and construction of roads were observed during the field investigations.

Villagers reported that earlier landslides were not very common but after the introduction of roads, this has become a common phenomenon. About 80 percent of the total landslides occurring now in the reserve are a result of road construction.

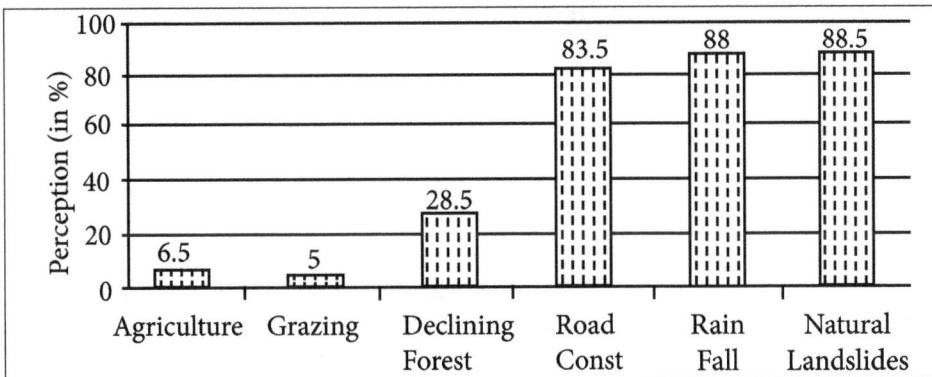

Peoples perceptions regarding causes of landslides.

According to field survey, most of villagers consider that road construction, excessive rainfall and seismicity are main causes of landslides. About 83.5%, 88% and 88.5% of the respondents reported road construction, excessive rainfall and seismicity respectively as the most common causes of landslide occurrence in the reserve. Whereas declining forest cover, grazing and agricultural land expansions are other causes of landslides.

Agriculture, grazing and deforestation are commonly regarded as the main causes of landslides in other Himalayan regions but these were reported as minor causes of landslides in the reserve, which is the result of ban on utilization of local resources in reserve. Road construction has been the prime cause of landslides in the region.

Most of the region is located in the periglacial and glacial environment. So, ideally, there should be much snowfall rather than the rainfall. But local people reported that snowfall has decreased and rainfall has increased in last few decades. This can be attributed to global warming due to which rainfall regions are shifting upwards taking place of snowfall in the reserve. Thus, excessive rainfall in the first half of monsoon saturates the bedrock and in latter half removes it in the form of landslides.

Road construction and rainfall have increased in the reserve thus landslides have also become very frequent. Survey results show that the landslides have been increasing rapidly in region since last three decades, as about 51 percent of the respondents reported the same, while 33 percent reported that landslides are increasing slowly and only 16 percent stated that trend of landslide is almost static.

Historical trend of major landslides in reserve has been constructed on the basis of literature and field survey. A total of 10 major landslides have been noticed in literature. There must be some more massive landslides in the region.

However, these are not properly documented due to the fact that most of the reserve is highly inaccessible to the outer world. Thus, most of the landslides of the reserve remain unregistered.

Table: History of Landslides in NDBR:

Year	Causes	Description
1939	-	Village was partially abandoned due to extensive rockfalls.
1978	Avalanches	Bamni village near Badrinath Puri was completely washed away.
1968	Related to Flash Floods	Rishi Ganga in Gar hwal was blocked upto a height of 40 m due to a slide at Reni village. The dam breached in 1970 caused extensive damage.
July 1970	Exceptionally Heavy Rainfall	Alaknanda river caused considerable loss of life among pilgrims. Many bridges. Houses and an entire village were washed away.
September 1970	Exceptionally Heavy Rainfall	Landslide and house collapses killed 223 people.

July 1970	landslide due to Dams	Floods in Rishi Ganga created 40m high blockade near village of Reni in U.P. Lake silted up by May. 1970 and eventually blockade breached in July, 1970.
September 1999	Earthquake-induced landslide	North of Joshimath

During field survey, about 40 landslides were noticed in reserve. Of the total landslides, about 31 were human-induced and 9 were natural. All the human-induced landslides have recent origin. Thus, it can be said that human activities have aggravated the problem of landslides only in recent past.

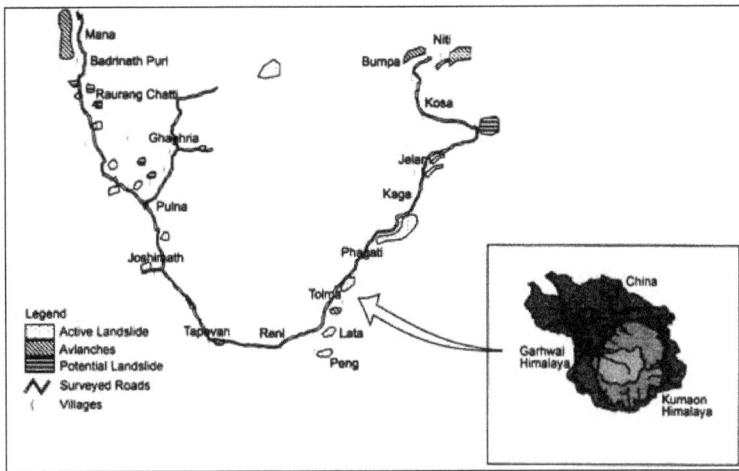

Landslides in Nanda Devi bio-sphere reserve.

Presently, intensity of landslide occurrence is high in NDBR as about 87 percent of the respondents indicated the same, whereas only 13% indicated moderate, low and very low intensity. People who reported moderate, low and very low intensity of landslides inhabit upper reaches of river valleys and practice seasonal migration.

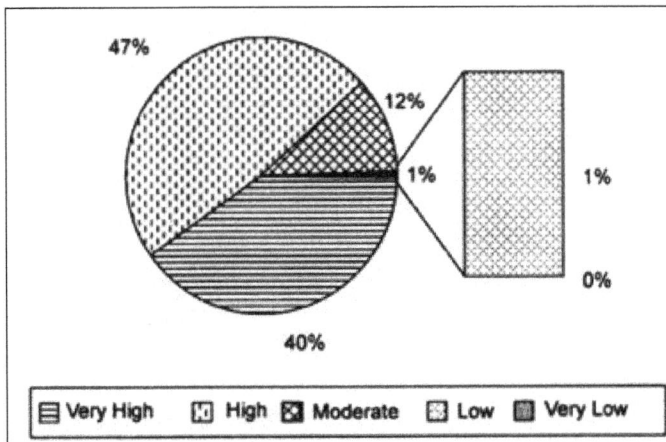

People's perceptions regarding intensity of landslides.

During monsoon, upper reaches receive less rainfall, thus landslides are not as common as in lower reaches of the river valley. During winters, when avalanches take place, inhabitants of upper reaches migrate to low altitude and do not witness cases of frequent landslides.

2.6.3 Soil Erosion and Desertification

Soil is naturally created when the small pieces of weathered rocks and minerals mix with the organic materials from decaying plants and animals. Soil creation is a slow process, taking many years. However, the soil that is created is constantly subjected to natural and manmade forces that disrupt it.

Soil erosion is a naturally occurring process that affects all the landforms. In agriculture, soil erosion refers to the wearing away of field's topsoil by the natural physical forces of water and wind or through the forces associated with farming activities such as tillage.

Causes of Soil Erosion

Erosion occurs when farming practices are not compatible with the fact that soil can be washed away or blown away. These practices are as follows:

- Inappropriate farming techniques such as deep ploughing of land 2 or 3 times per year to produce the annual crops.

- Overstocking and overgrazing.

- Planting the crops down the contour instead of along it.

- Lack of crop rotation.

Erosion, whether it is by water, wind or tillage, involves three distinct actions such as soil detachment, movement and deposition. Topsoil, which is high in organic matter, fertility and soil life is relocated elsewhere "on-site" where it builds up over time or is carried "off-site" where it fills in the drainage channels. Soil erosion reduces the cropland productivity and contributes to the pollution of adjacent wetlands, watercourses and lakes.

Soil erosion can be a slow process that continues relatively unnoticed or can occur at an alarming rate, causing serious loss of topsoil. Soil compaction, loss of soil structure, low organic matter, salinization, poor internal drainage and soil acidity problems are other serious soil degradation conditions that can accelerate the soil erosion process.

Soil erosion is defined as the wearing away of topsoil. Topsoil is the top layer of soil and is the most fertile because it contains the most organic, nutrient-rich materials. Hence, this is the layer that farmers wants to protect for growing their crops and ranchers want to protect for growing grasses for their cattle to graze on.

One of the main causes of soil erosion is water erosion, which is the loss of topsoil due to water. Raindrops fall directly on the topsoil. The impact of the raindrops loosens the material bonding it together, allowing small fragments to detach. If the rainfall continues, water gathers on the ground, causing water flow on the land surface which is termed as surface water runoff. This runoff carries the detached soil materials away and deposits them elsewhere.

There are some conditions that can accentuate the surface water runoff and therefore, soil erosion. For example, if the land is sloped, there is a greater potential for soil erosion due to gravity which pulls the water and soil materials down the slope.

Effects of Soil Erosion

The loss of natural nutrients and possible fertilizers directly affect the crop growth, emergence and yield. Seeds can be disturbed or removed and pesticides can be carried off. The soil structure, quality, texture and stability are also affected, which in turn affects the holding capacity of the soil.

Eroded soil can inhibit the growth of seeds, bury seedlings, contribute to road damage and even contaminate the water sources and recreational areas.

Desertification

Desertification is a progressive destruction or degradation of arid or semi-arid lands. It is also a form of land degradation. Desertification leads to the conversion of range lands or irrigated croplands to desert like conditions in which agricultural productivity falls. Desertification is characterized by denegation, depletion of ground water, desalination and soil erosion.

Causes of Desertification

- Deforestation: The process of deserting and degrading a forest land initiates a desert. If there is no vegetation to hold back the rain water, soil cannot soak and ground water levels do not increases. This also increases, soil erosion, loss of fertility.

- Overgrazing: The increase in cattle population heavily graze the grass land or forests and as a result denude the land area. The denuded land becomes dry, loose and more prone to soil erosion and leads to desert.

- Pollution: Excessive use of fertilizers and pesticides and disposal of toxic water into the land also leads to desertification.

Harmful Effects of Desertification

- Around 80% of the productive land in the arid and semi-arid regions are converted into desert.

- Around 600 million people are threatened by desertification.

Chapter 3

Ecosystems and Biodiversity

3.1   Concept of an Ecosystem

A group of organisms interacting among themselves and with the environment is known as ecosystem. Thus, an ecosystem is a community of different species interacting with one another and with their non-living environment exchanging energy and matter.

Example: Animals cannot synthesize their food directly but depends on the plants either directly or indirectly.

Few examples of ecosystem are pond, forest, estuary and a grassland.

Primary Types of Ecosystems

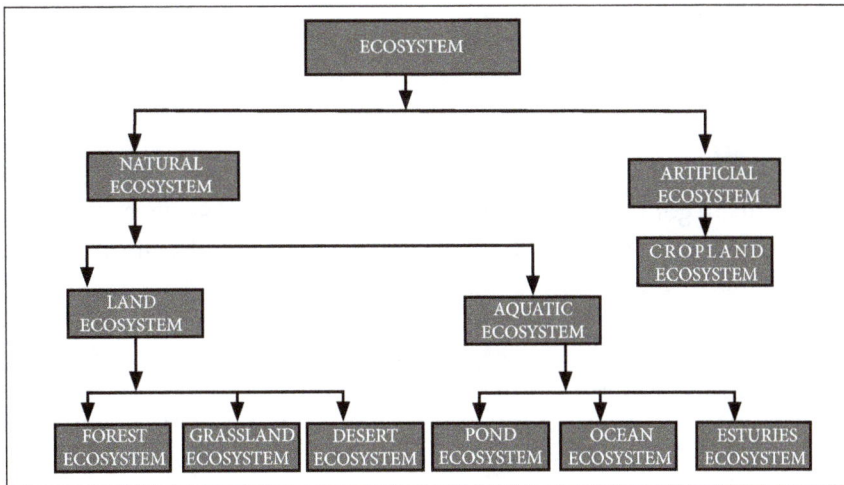

Natural Ecosystems

Natural ecosystems may be terrestrial or aquatic. A natural ecosystem is a biological environment that is found in nature rather than created or altered by man.

Artificial Ecosystems

Humans have modified some ecosystems for their own benefit. These are the artificial ecosystems. They can be terrestrial or aquatic.

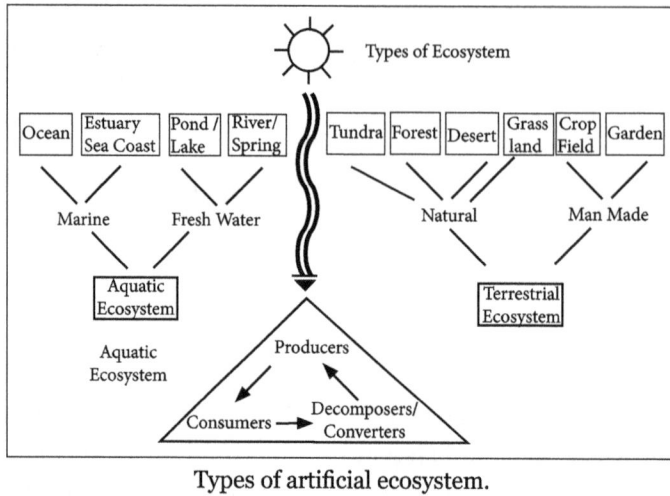

Types of artificial ecosystem.

Types of Natural Ecosystems

There are two main types of natural ecosystems. They are:

- Aquatic natural ecosystem.

- Terrestrial natural ecosystem.

In aquatic ecosystems or ganisms interact with water. In terrestrial ecosystems organisms interact with land.

Aquatic Ecosystems

Aquatic ecosystems in general covers up to 71% of the earth's surface. As a type, aquatic ecosystems can be classified again into three varieties, defined by the kind of water with which the organisms interact.

Freshwater

This type of ecosystem includes rivers, lakes, streams, ponds and wetlands and makes up smallest percent of the earth's aquatic ecosystem.

Transitional Communities

These are places where freshwater and salt water comes together, including estuaries and wetlands.

Marine

More than 70% of the earth is covered by salt water or marine ecosystems. These include shorelines, coral reefs and open ocean.

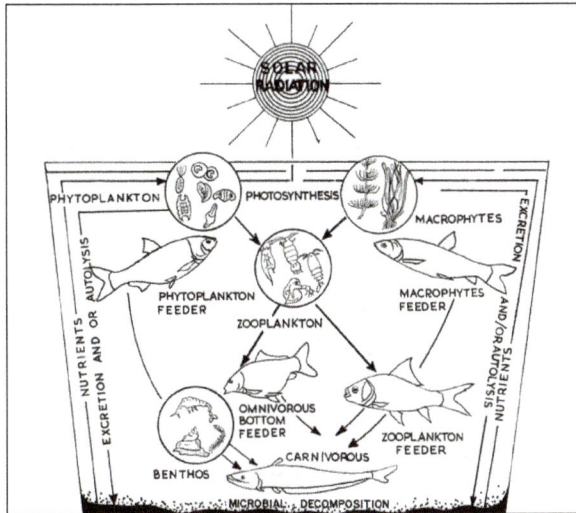

Pond as an ecosystem.

Terrestrial Ecosystems

Terrestrial ecosystems are classified by the type of land or terrestrial area. Mountains, forests, deserts and the grasslands are types of terrestrial ecosystems:

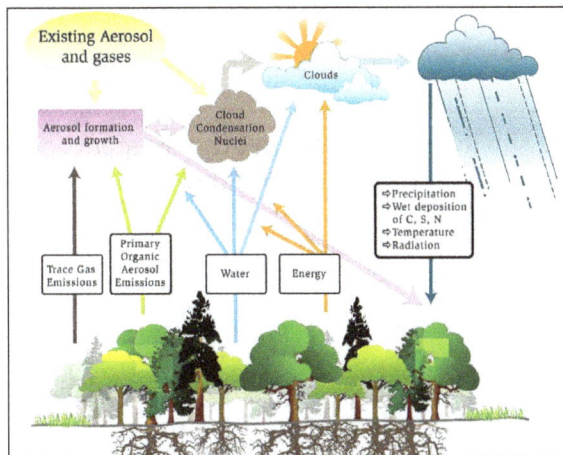

Terrestrial ecosystems.

- Forest: These ecosystem features dense tree population and include tropical rain forests.

- Desert: Deserts receive less than 25 cm of rainfall per year.

- Grassland: These ecosystem includes tropical savannas, temperate prairies and arctic tundra.

- Mountain: Mountain ecosystem includes steep elevation changes between meadows, ravines and peaks.

3.1.1 Structure and Function

Structure of an Ecosystem

The term structure refers to various components. So the structure of an ecosystem explains the relationship between the biotic and abiotic components.

An ecosystem has two major components:

- Biotic components.

- Abiotic components.

Components of ecosystem.

Biotic Components

The living organisms or the living members in an ecosystem collectively form its community called biotic components or biotic community.

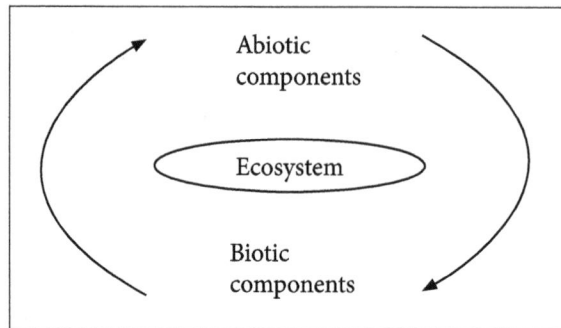

Example: Plants, animals and micro-organisms.

Classification of Biotic Components:

The members of the biotic components of an ecosystem are grouped into three groups based on how do they get their food;

- Producers (Plants).

- Consumers (Animals).

- Decomposers (Micro-organisms).

Grass →	Rat →	Cat →	Tiger
Producers	Herbivores	Primary Carnivores	Secondary Carnivores

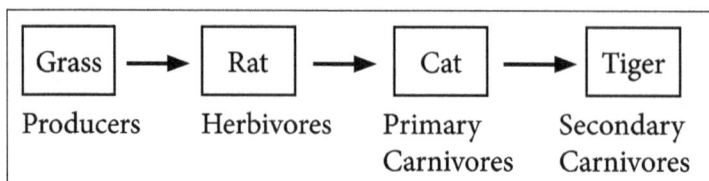

Biotic components of an ecosystem.

Producers

Producers synthesize their food themselves through the process of photosynthesis. Plants are known as producers, because they produce their own food. They do this by using carbon dioxide from the air, light energy from the sun and water from the soil to produce food in the form of glucose or sugar. This process is called photosynthesis.

Example: All green plants and trees.

Consumers

Consumers are organisms, which cannot prepare their own food and depend directly or indirectly on the producers. Animals are called consumers, because they cannot make their own food, so they need to consume plants or animals. Humans are also omnivores.

Example:

- Plant eating species like insects, rabbit, goat, etc.

- Animals eating species like lion, tiger, etc.

Group of Consumers:

There are three groups of consumers. They are;

- Animals that eat only plants.

- Animals that eat only animals.

- Animals that eat both plants and animals.

Classification of Consumers:

Consumers are further classified into;

1. Primary Consumers

Primary consumers are also called herbivores; they directly depend on the plants for their food. So they are known as plant eaters.

Example: Insects, rat and goat.

2. Secondary Consumers

Secondary consumers are the primary carnivores, they feed on primary consumers. They directly depend on the herbivores for their food.

Example: Frog, cat, etc.

3. Tertiary Consumers

Tertiary consumers are secondary carnivores, they feed on secondary consumers. They directly depend on the primary carnivores for their food.

Example: Tigers, lions, etc.

Decomposers

Decomposers attack the dead bodies of producers and consumer and decompose them into simpler compounds. During decomposition, inorganic nutrients are released. These inorganic nutrients together with other organic substances are then utilized by the producers for the synthesis of their own food.

Example: Micro-organisms like bacteria and fungi.

Bacteria and fungi are decomposers. They eat decaying matter, dead animals and plants and in the process they break them down and decompose them. Then they release nutrients and mineral salts back into the soil which will be used by plants.

Abiotic Components

The non-living components of the ecosystem collectively form a community called abiotic components or abiotic community.

Example: Climate, soil, water and air.

1. Physical Components:

They include the climate, energy, raw materials and living space that the biological community needs. They are useful for the growth and maintenance of its member.

Example: Air, water, soil, sunlight, etc.

2. Chemical Components:

They are the sources of essential nutrients.

Example:

- Organic substances: Protein, carbohydrates, lipids, etc.
- Inorganic substances: All micro and macro nutrients and few other elements.

Functions of an Ecosystem

The function of an ecosystem is to allow flow of energy and cycling of nutrients.

Types of Functions

Functions of an ecosystem are of three types. They are as follows:

- Primary function: The primary function of all ecosystems is the manufacture of starch.

- Secondary function: The secondary function of all ecosystem is distributing energy in the form of food to all consumers.

- Tertiary function: All living systems die at a particular stage. These dead systems are decomposed to initiate the third function of ecosystem known as "cycling".

Functions of an Ecosystem

Major functions of an ecosystem which ensures its stability are as follows:

- It regulates flow rate of biological energy including production and respiration rates.

- It regulates flow rates of materials in terms of mass balance as nutrients or materials cycles.

- It fixes limit of tolerance for each organisms in an ecosystem because each organism has certain limits of tolerance towards various factors of environment and only within specified limits, the organisms survive.

- It regulates modification of environment because the environment is modified by the organisms according to their needs. The carrying capacity of the ecosystem determines the size of organisms population that can survive in the particular ecosystem.

- It regulates species diversity because the 'nature strives for greater diversity i.e., greater variety of organisms in a system, which leads to its stability.

- It controls alteration of any one component because it is holocoenotic in which one component effects the other components.

3.2   Energy Flow

Energy is defined as the ability to do work. The Einstein equation $E = mc^2$ implies that matter and energy are interchangeable. Tremendous energy is stored in atoms that collectively form matter. Biological activity involves the utilization of energy.

It comes ultimately from the sun and is transformed from the radiant energy to chemical energy by photosynthesis and chemical form to mechanical form or heat by cellular metabolism.

There are two kinds of energy, i.e., potential energy and kinetic energy. Energy at rest is called the potential energy, where in motion/work, it is called kinetic energy. The behavior of energy can be described by the laws of thermodynamics.

On earth, three sources of energy that account for all the work in an ecosystem viz. gravitation, internal forces within the earth and solar radiation. The solar radiation which originates from the sun is the source of energy for life. Only about 150 millionth of the sun's tremendous energy reaches the earth's outer atmosphere at a constant rate, which is called as solar flux.

The amount of radiant energy that cross a unit area or surface unit of time is estimated to be 2 calories per square centimeter per minute for a total income of 13×10^{23} cal/year.

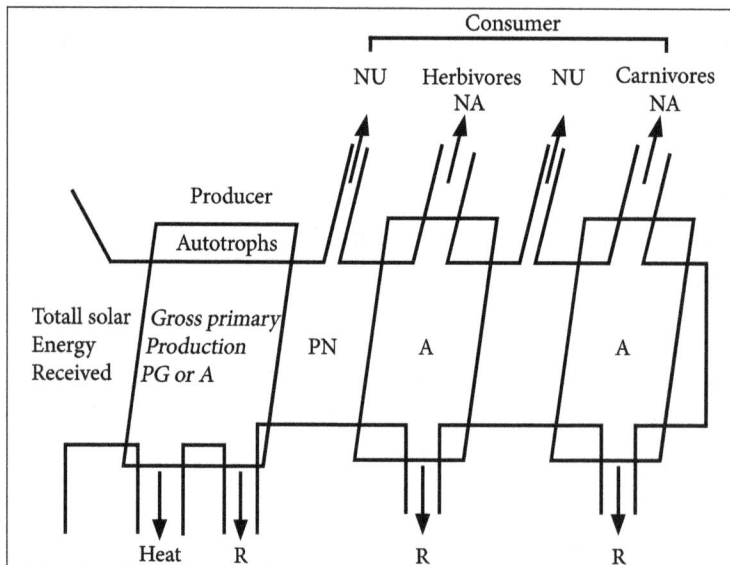

Energy flow in an ecosystem.

Energy flow through a trophic level:

> = Assimilation (A) at that level.

> = Production of biomass 'P' + Respiration 'R'.

> I = Total energy input.

> PG = Gross primary production.

> A = Total assimilation.

> PN = Net primary production.

P = Secondary production,

NU = Energy not used (stored or exported),

NA = Energy not assimilated (ejected),

R = Respiration.

Solar energy in kCal/m²/day	3000	1500	15	1.5	0.3
	L	LA	PN	P_2	P_3

Ecosystems are real like a pond or a field, a forest, an ocean or even an aquarium. In spite of great diversity in types of actual ecosystems from small to large, terrestrial to aquatic etc. Energy flow in the ecosystem keeps it going. Its flow is unidirectional and is from producers to herbivores to carnivores. It cannot occur in the reverse direction.

Also the amount of energy decreases with successive trophic levels (trophic = nourishment). It moves in the ecosystem starting from autotroph (the producer level, i.e., first trophic level) to heterotrophs including plant eaters or herbivores (primary consumer i.e., second trophic level), primary carnivores which eat the herbivores (secondary consumer i.e., third trophic level) and secondary carnivores (tertiary consumer i.e., fourth trophic level).

It means energy is transferred from one trophic level to the other in succession in the form of a chain called food chain. The amount of energy decreases with successive trophic levels.

Green plants use only up to 5% of the total solar radiation and the rest unutilized energy is dissipated as heat. An estimate shows that the energy actually used by the herbivores is only 10% of the gross productivity of producers. Similarly herbivores use part of this energy for growth and maintenance.

The rest is left as fecal matter or dead organic matter and the rest is used to support carnivore population. In food chain another animals of the chain get only 10% the energy.

3.2.1 Ecological Succession

The progressive replacement of one community by another life along the development of the stable community in a particular area is called ecological succession.

Stages of ecological succession:

- Pioneer community: The first group of organism which establish their community in the area is termed as "Pioneer community."

- Seres or seral stage: The various developmental stages of the community is called as "seres".

- Community: It is the group of plants or animals living in an area.

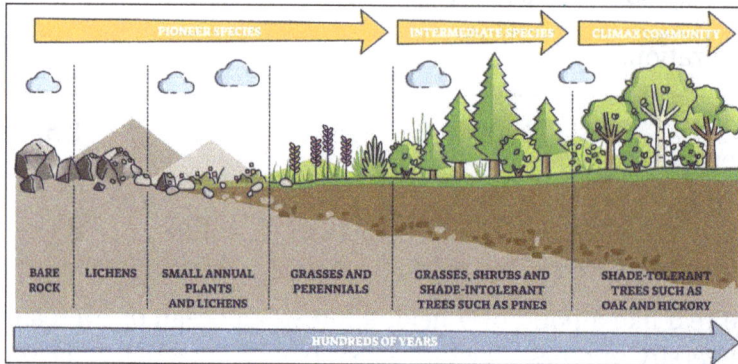

Ecological succession.

Types of Ecological Succession

Ecologists recognize two types of ecological succession based on the conditions present at the beginning of the process such as:

1. Primary Succession

It involves the gradual establishment of biotic communities on a lifeless ground:

- Hydrarch (or) Hydrosere: Establishment starts in a watery area like pond or lake.

- Xerarch: Establishment starts in a dry area like deserts and rocks.

2. Secondary Succession

It involves the establishment of biotic communities in an area where some type of biotic community is already present.

Process of Ecological Succession

The process of ecological succession can be explained in following steps:

1. Nudation

It is the development of a bare area without any life form.

2. Invasion

A biological invader is a species of animal, plant or microorganism which most usually transported intentionally by man and spreads into new territories some distance from its home territory. If one looks at long time scales then invasions are probably a frequent phenomenon. However, human activity has increased the frequency of invasions dramatically by disrupting biogeographic barriers and increasing exchanges:

- Migration: Migration of seeds is brought about by water, wind or birds.

- Establishment: The seeds then germinate and grow on the land and establishes their pioneer communities.

3. Competition

As the number of individual species grows, there arises competition with the same species and between different species for water, space and nutrients.

4. Reaction

The living organisms take water, nutrients and grows and modifies the environment is known as reaction. This modification becomes unsuitable for the existing species and also favors some new species, which replace the existing species.

5. Stabilization

It leads to stable community, which is in equilibrium with the environment.

3.2.2 Ecological Pyramids

Graphical representation of structure and function of tropic levels of an ecosystem, starting with producers at the bottom and each successive tropic level forming the apex is known as an ecological pyramids.

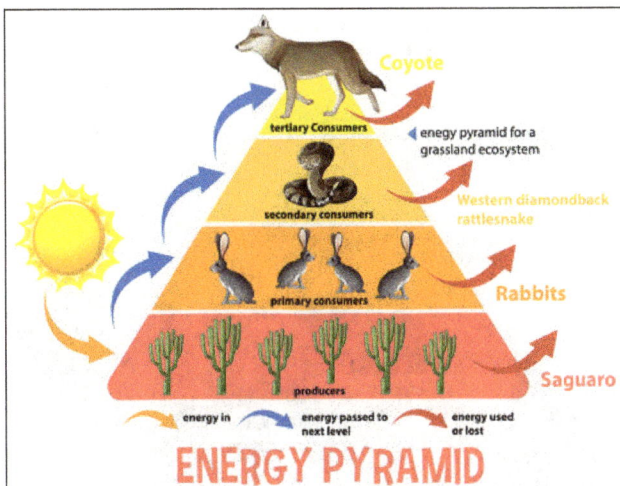

Ecological pyramids.

In the food chain starting from producers to consumers, there is a regular decrease in the properties. Since some energy is lost as heat in each tropic level, it becomes progressively smaller near the top.

Types of Ecological Pyramids

Ecological pyramids are of three types:

- Pyramid of number.

- Pyramid of energy.

- Pyramid of biomass.

1. Pyramid of Numbers

It represents the number of individual organisms present in each tropic levels.

Example: A grassland ecosystem.

The producers in the grasslands are grasses, which are small in size and large in numbers. So the producers occupy lower tropic levels.

The primary consumers are rats, which occupy the IInd tropic level. Since the number of rats are lower when compared to the grasses, the size of which is lower.

The secondary consumers are snakes, which occupy the IIIrd tropic levels. Since the number of snakes are lowers when compared to the rats, the size of which is lower.

The tertiary consumers are eagles, which occupy the next tropic level. The size of the last tropic level is lowest.

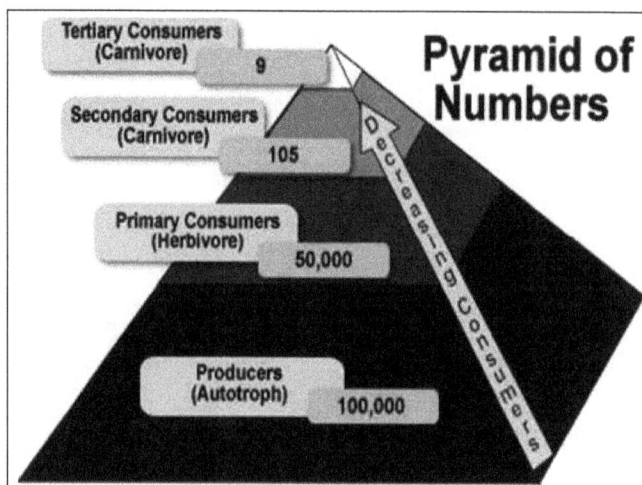

Pyramid of numbers.

2. Pyramids of Energy

This represents the amount of energy present in each tropic level. The rate of energy flow and the productivity at each successive tropical level is shown.

At every successive tropic level, there is a heavy loss of energy in the form of heat. Thus at each higher tropic level only 10% of the energy is transferred. Hence, there is a sharp decreases in energy at each and every successive tropic level as we move from producers to top carnivores.

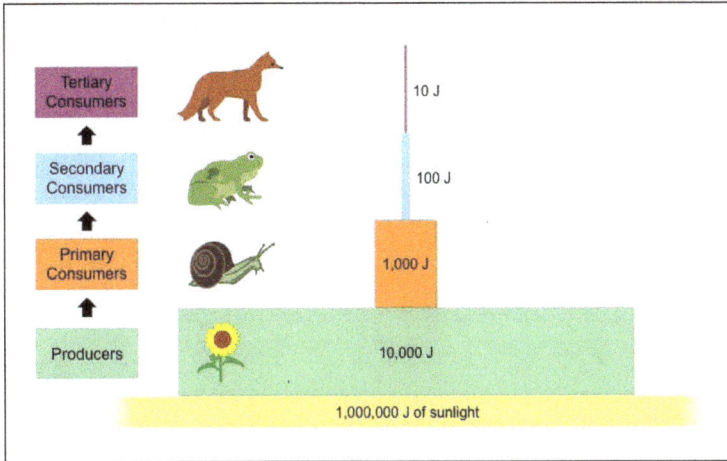

Pyramid of energy.

3. Pyramids of Biomass

It represents the total amount of biomass present in each tropic levels.

Example: A forest ecosystem.

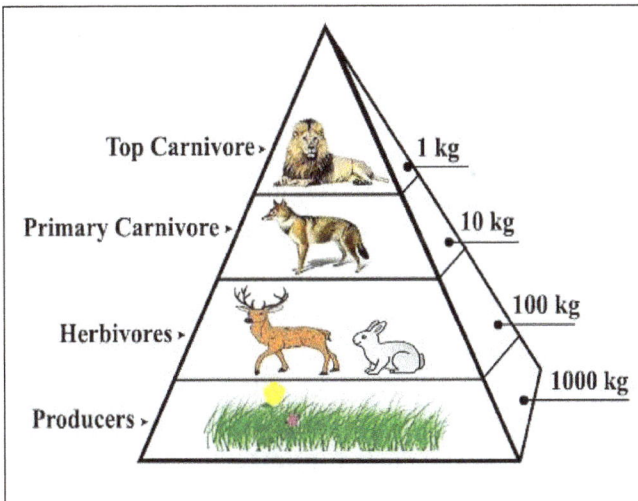

Energy of Biomass.

The above figure shows that there is a decrease in the biomass from the lower tropic level to the higher tropic level. This is because the producers (trees) are maximum in the forest, which contribute a huge biomass.

The next tropic levels are herbivores (insects, birds) and carnivores (snakes, foxes). The top of the tropic level contains few tertiary consumers (lions and tigers), the biomass of which is very low.

3.3   Forest, Grassland, Desert and Aquatic Ecosystems

Forest Ecosystem

The entire assemblage of organisms together with their environmental substrate, interacting inside a defined boundary. Forests and woodlands occupy about 38% of the Earth's surface and they are more productive and have greater biodiversity than other types of terrestrial vegetation.

The ecological benefits:

- Production of oxygen: During photosynthesis trees produce oxygen which is essential for life on earth.

- Reducing global warming: The main greenhouse gas carbon dioxide (CO_2) is absorbed by the trees (forests).

- The trees absorb the main greenhouse gas carbon dioxide (CO_2), which is a raw material for photosynthesis. Thus the problem of global warming caused by greenhouse gas CO_2 is reduced.

- Soil conservation: Roots of trees bind the soil tightly and prevents soil erosion. They also act as wind breaks.

- Regulation of hydro logical cycle: Watersheds in forest act like giant sponges which absorb rainfall, slow down the runoff and slowly release the water for recharge of springs.

- Pollution modulators: Forests can absorb many toxic gases and noises and helps in preventing air and noise pollution.

- Wildlife Habitat: Forests are the homes of millions of wild animals and plants.

Characteristics of Forest Ecosystems

Forests are characterized by warm temperature and adequate rainfall, which make the generation of number of ponds, lakes etc.:

- The forest maintains climate and rainfall.

- The forest support many wild animals and protect biodiversity.

- The soil is rich in organic matter and nutrients, which support the growth of trees.

- Since penetration of light is so poor, the conversion of organic matter into nutrients is very fast.

Types of Forest Ecosystem

Depending upon the climate conditions, forest can be classified into the following types:

- Tropical deciduous forests.

- Tropical rain forests.

- Temperate rain forests.

- Tropical scrub forests.

- Temperate deciduous forests.

Features of Different Types of Forests

Tropical Rain Forests:

They are found near the equator. They are characterized by high temperature. They have broad leaf trees like teak and sandal and the animals like lion, tiger and monkey.

Tropical Deciduous Forests:

They are found little away from the equator. They are characterized by a warm climate and rain is only during monsoon. They have different types of deciduous trees like maple, oak and animals like deer, fox, rabbit and rat.

Tropical Scrub Forests:

These are characterized by a dry, climate for longer time. They have small deciduous trees and shrubs and animals like deer, fox, etc.

Temperate Rain Forests:

They are found in temperate areas with adequate rainfall. They are characterized by coniferous trees like pines, red wood, firs, etc. and animals like squirrels, fox, cats, bear, etc.

Temperate Deciduous Forests:

They are found in areas with moderate temperatures. They have major trees including broad leaf deciduous trees like oak, hickory and animals like deer, fox, bear, etc.

Structure and Function of Forest Ecosystem

Abiotic Components

Examples: Climatic factors and minerals.

The abiotic components are inorganic and organic substances found in the soil and atmosphere. In addition to minerals, the occurrence of litter is characteristic features of majority of forests.

Biotic Components

1. Producers

Examples: Trees, Shrubs and ground vegetation.

The plants absorb sunlight and produce food through photosynthesis.

2. Consumers

i. Primary Consumers:

Examples: Ants, flies, insects, mice, deer, squirrels.

They directly depend on the plants for their food.

ii. Secondary Consumers:

Examples: Snakes, birds, fox.

They directly depends on the herbivores for their food.

iii. Tertiary Consumers:

Examples: Animals like tiger, lion, etc.

They depend on the primary carnivores for their food.

3. Decomposers

Examples: Bacteria and fungi.

They decompose the dead plant and animal matter. Rate of decomposition in tropical and subtropical forests is more rapid than in the temperate forests.

Northern Coniferous Forests

Below the tundra region, both in North America and Eurasia, lie the northern coniferous forests. The most prominent vegetation is the needle-leaved evergreen trees especially the spruces, firs and pines. A dense shade results in a poor undergrowth. As a result of the evergreen nature of the forests, productivity is fairly high, although there is low temperature throughout half of the year.

The animals found in these forests are moose, snowshoe hare, grouse, squirrels, crossbills etc. The coniferous forests often experience bark beetle outbreaks, thus paving the way for succession in the ecosystem.

Moist Temperate Coniferous Forests

Here, in this type of ecosystem, temperatures are a bit higher than the northern coniferous type and here humidity is very high because of the dense fog which often substitutes for reduced precipitation in certain areas. As water is not a general limiting factor, these regions are called the temperature rain forests also. Rainfall ranges from 30-150 inches.

Trees such as western hemlock, Douglas fir, redwoods and Sitka spruces are the dominant species. Wherever there is any penetration of light, the understorey is well developed.

Temperate Deciduous Forests

Deciduous forest communities occupy areas with abundant, evenly distributed rainfall (30-60 inches) and moderate temperatures. It covers most of Europe, eastern North America, part of Japan, etc. Hence there are more isolated forest regions and species composition differs immensely. As leaves wither from the trees at least during a certain period in every year, the contrast between every season is great.

These forests represent one of the most important biotic regions of the world, because Europeans and North Americans (settlers) have modified these areas and now prime forests have been modified by the human communities.

Beeches, maples, oaks, hickories, chestnuts and other trees are the most common climax vegetation of these forests.

Broad-leaved evergreen subtropical forests: Where the moisture remains high and temperature differences between winter and summer are narrowed down, the temperature of deciduous forest gives way to the broad-leaved evergreen forest climate. Trees as varied as northerly oaks, to the more tropical strangler fig, wild tamarind, etc. are present in these type of forests.

Tropical Rain-forests

The variety of life reaches its diverse best in tropical rainforests. These forests are present in the tropical regions and low altitudes. Rainfall exceeds 80-90 inches per year and is distributed all the year round. The forests of South America, the Amazon basin, are the largest and the most contiguous of all the rain forests. The biodiversity of the forests is tremendous.

Any tropical rain forest is highly stratified with generally three to four layers or 'stores'. They are:

- Scattered very large trees that project above the canopy.

- Canopy layer which forms a continuous carpet-like layer about 80-100 feet tall.

- An understorey, which is present only where there is a break in the canopy leading to the sun's light to reach the bottom.

If rainfall becomes less during the dry season, another subtype of forests called the semi-evergreen type results. Shrub and herb layers often having ferns and palms are present in the understorey. Plants like epiphytes grow on the massive tree trunks and vines and lianas vie for space and light.

Most of the species of animals present here are arboreal and much are herpetological forms and birds. Insect life is also very diverse. The animals like geckos, snakes, frogs, parakeets, hornbills, cotingas, monkeys and predators like clouded leopard are present.

Tropical scrub and deciduous forests. Where moisture conditions are intermediate between desert and Savannah on the one hand and rain forest on the other, tropical scrub or thorn forests and tropical deciduous forests are predominant.

The tropical Asian forests are mostly of the tropical deciduous type. Wet and dry spells of equal length alternate and hence the contrast between the seasons are very conspicuous.

The scrubland habitat supports thorny trees, shrubs and the main climatic factor is the imperfect distribution of fairly good total rainfall.

3.3.1 Grassland Ecosystem

Grasslands occupy about 19% of the earth's surface. The major grassland ecosystems of the world are the great plains of Canada and United States, Argentina to Brazil and Asia to Central Asia.

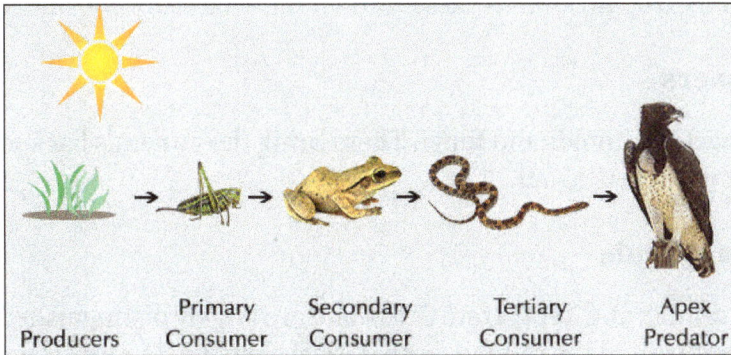

Food chain in grassland ecosystem.

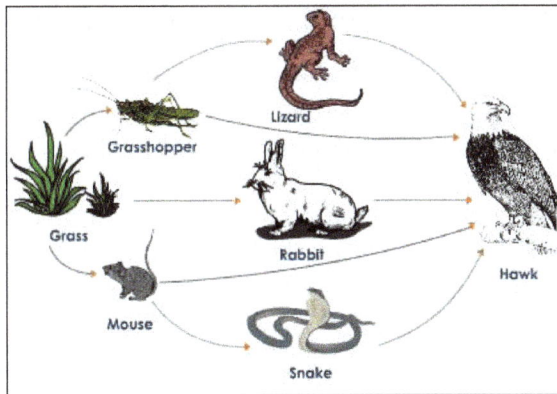

Food web in a grassland ecosystem.

The various components of a grassland ecosystem are as follows:

1. Abiotic Substances

These include the nutrients present in the soil and the aerial environment. The elements required by plants are hydrogen, oxygen, nitrogen, phosphorous and sulphur. These are supplied by the soil and air in the form of CO_2, water, nitrates, phosphates and sulphites. In addition to these some trace elements are also present in the soil.

2. Primary Producers

These are mainly grasses of the family, Graminae, large variety of herbs, some shrubs and scattered trees.

3. Consumers

Herbivores such as grazing mammals, insects, some termites and millipedes are primary consumers.

The animals like fox, jackals, snakes, frogs, lizard, birds etc., are the carnivores feeding on herbivores. These are the secondary consumers of the grassland ecosystem. Hawks occupy the tertiary trophic level as these feed on the secondary consumers.

4. Decomposers

These include bacteria, molds and fungi. These bring the minerals back to the soil to be available to the producers again.

Types of Grasslands

The tropical Savannah and Temperate Grassland are largely distinguished by differences in the temperature and the rainfall, both critical elements to a grassland's formation. An area that receives very little rain becomes a desert, an area that receives significant amounts of rain often develops into forest. Grasslands hang somewhere in the balance.

Tropical Savannah's: It is found in Africa, Australia, South America and Indonesia, stay warm all year. They receive 50 to 130 cm during the rainy season (6 to 8 months) and endure drought for the remainder of the year.

Plant and animal species vary greatly across Savannah, curbed by differences in climate, but much of the Savannah is characterized by thin soil where only grasses and flowering plants can grow. Like Canada's grasslands, this ecosystem supports an astonishing diversity of species, the African savanna, for example, is home to some of the world's most iconic mammals, including giraffes, zebras and lions.

Temperate Grasslands: It include Canadian grassland ecosystems, are also found around the globe. The plant and animal species in temperate grasslands are shaped by less rainfall (25 to 90 cm) and cycle through a greater range of seasonal temperatures. Many temperate grassland animals, which must adapt to dry, windy conditions, are recognizable to Canadians. Grazing species like antelope and elk, burrowing animals like prairie dogs and badgers and predators like snakes and coyotes.

The dramatic contours of Canada's grasslands are the result of glacial movement and melting ice, which shaped this landscape over the last two hundred million years. Grasslands National Park, for example, boasts glacial meltwater channels that feature plateaus, coulees, buttes that rise abruptly at horizon and layers of rock formation that hold fossilized secrets from 80 million years ago.

Characteristics of the Grassland

Precipitation

Grasslands make up 25 percent of the Earth's land surface and dominate in regions with limited rainfall, which prevents forest growth. This is the result of nearby mountain ranges that cause rain shadows over adjacent open range lands. Usually, grasslands have not only limited but also unpredictable rainfall and droughts are common.

Where rainfall is even less, deserts will form. Savannas, on average, receive roughly 76 to 101 centimeters (30 to 40 inches) of rain per year, but steppes only average 25 to 51 cm (10 to 20 inches) per year. Prairies tend to be intermediate between savannas and steppes with 51 to 89 centimeters (20 to 35 inches) per year.

Temperature

It very much more in temperate grasslands than they do in savannas. Savannas are located in warm climates with average annual temperatures that only vary between 21 and 26 degrees Celsius (70 and 78 degrees Fahrenheit).

They usually have only two seasons, a wet and a dry season. Temperate grasslands are characterized by hot summers where temperature can exceed 38 degrees Celsius (100 degrees Fahrenheit) and cold winters that can drop below negative 40 degrees Celsius (negative 40 degrees Fahrenheit).

Fire

Fires are an important grassland characteristic. Regular fires promote the growth of native grasses but limit the growth of trees. Native grasses have deeper root systems that can survive fires, but invasive plants tend to have shallower roots and succumb to fires.

Development has curtailed the number and extent of grassland fires and the lack of seasonal fires threatens the health of the world's grasslands. As of 2013, only 5% of the world's grasslands were protected and properly maintained and they remain the most endangered biome in the world.

Flora and Fauna

Savannas are home to some of the largest mammals on the planet like elephants, giraffes, rhinos, lions and zebras. Temperate grasslands are also home to large mammals, particularly bison and horses, medium-sized mammals like deer, antelope and coyotes, as well as small mammals such as mice and jack rabbits. The type of grasses that grow depend upon the amount of rainfall. Shorter steppe grasses often consist of buffalo grass and savanna grasses will contain taller grasses like bluestem and rye.

3.3.2 Desert Ecosystem

One can find at least one desert on every continent except Europe and Antarctica. In order to be considered as a desert, it must receive less than 10 inches of water per year.

There are plenty of differences between the deserts of world. Some deserts are made of fine, red sand, others consist of sand mixed with pebbles and rocks. The desert sand started out as rock but after some years, weathering by wind and water has created dunes in the deserts. These sands are mostly minerals and sometimes oil can be found hidden deep within the rocks.

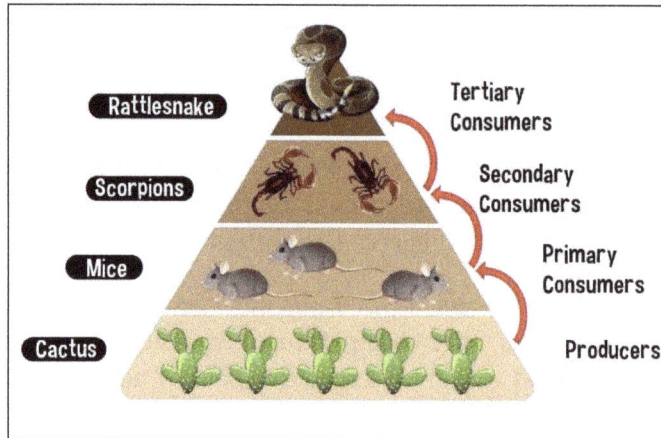

Desert Ecosystem.

Types and Characteristic Features

One can find at least one desert on every continent except Europe and Antarctica. Each desert is different in some way, but they all have one thing in common. In order for an area of land to be considered a desert, it must receive less than 10 inches of water a year.

Clouds are scarce in these regions and we all know that without the clouds, there can't be rain, snow or any other precipitation. But clouds also serve another purpose, they block out some of Sun. The desert gets mighty hot during the day because the Sun beats down on the sand. At night, desert gets very cold, because there are not clouds around to keep the heat from escaping to the atmosphere.

There are plenty of differences between the deserts of the world. Some deserts are made of very fine, red sand, others consist of sand mixed with pebbles and rocks. The desert sand started out as rock, but years of weathering by wind and water has created dunes in the deserts. These sands are mostly minerals and sometimes oil can be found hidden deep within the rocks.

Structure and Function

The various components of a desert ecosystem are:

1. Abiotic Component

It includes the nutrients present in the soil and in aerial environment. The characteristic feature of the abiotic component is the lack of organic matter in the soil and scarcity of water.

2. Biotic Component

The various biotic components representing three functional groups are:

i. Producer Organisms

The producers are mainly shrubs or bushes, some grasses and a few trees. Surprisingly, there are many species of plants that can survive in the desert. Most of them are succulents, which mean they store water. Others have the seeds that lay dormant until a rain awakens them. Regardless, these plants find a way to get water and protect themselves from the heat.

The most famous desert plant is cactus. There are many species of cacti. The saguaro cactus is the tall, pole shaped cactus. The saguaro can grow to a height of 40 feet. It can hold several tons of water inside its soft tissue. Like all other cacti, saguaro has a thick, waxy layer that protects it from the Sun.

Other succulents include the desert rose and the living rock. This strange plant looks like a spiny rock. Its disguise protects it from predators. The welwitschia is a type of weird looking plant. It has two long leaves and a big root. This plant is actually a type of tree and it can live for thousands of years.

There are many other kinds of desert plants. Some of them have beautiful flowers, others have thorns and deadly poisons. Even in worst conditions, these plants continue to thrive.

ii. Consumers

These include animals such as the insects and reptiles. Besides them, some rodents, birds and some mammalian vertebrates are also found.

Desert Insects and Arachnids

There are plenty of insects in the desert. One of the most common and destructive pests is the locust. Locust is a special type of grasshopper. They travel from place to place, eating all the vegetation they find. It can destroy many crops in a single day.

Not all desert insects are bad, though. The yucca moth is very important to the yucca plant, because it carries pollen from the flower to the stigma. The darkling beetle has a hard, white, wing case that reflects the Sun's energy. This allows the bug to look for food during the day.

There are also several species of ants in desert. The harvester ants gather seeds and store them for use during the dry season. And the honey pot ants have a very weird habit. Some members of the colony eat large amounts of sugar, so much that their abdomens get too large for them to move. The rest of the colony feeds off this sugar.

There are also arachnids in the desert. The spiders are the most notable arachnids, but scorpions also belong in this group. Some species of scorpions have poison in their sharp tails. They sting their predators and their prey with the piercing tip.

Desert Reptiles

Reptiles are some of the most interesting creatures of the desert. Reptiles can withstand the extreme temperatures because they can control their body temperatures very easily. We can put most of the desert reptiles into one of two categories: snakes and lizards.

Many species of rattlesnakes can be found in the desert. Rattlesnakes have a noisy rattle they use to warn enemies to stay away. If the predator is not careful, the rattlesnake will strike, injecting venom with its sharp fangs. Other desert snakes include cobra, king snake and the hognose.

Lizards make up the second category of desert reptiles. They are probably the most bizarre looking animals in the desert. While some change colors and have sharp scales for defense, others change their appearance to look more threatening.

One such creature is the frilled lizard. When enemies are near, lizard opens its mouth, unveiling a wide frill. This makes the lizard look bigger and scarier. The shingle back has a tail with the same shape as its head. When a predator bites at the tail, the shingle back turns around and bites back. There are only two venomous lizards in the world and one of them is the gila monster. It has a very painful bite.

Desert Birds

Like the other inhabitants of desert, birds come up with interesting ways to survive in the harsh climate. The sand grouse has special feathers that soak up water. It can then carry the water to its young trapped in the nest.

Other birds, like gila woodpecker, depend on the giant saguaro as its home. This woodpecker hollows out a hole in the cactus for a nest. The cool, damp inside is safe for the babies.

The roadrunner is probably the most well-known desert bird. The roadrunners are so named because they prefer to run rather than fly. Ostriches also prefer to use their feet. Even the young depend on walking to find food and water. The galah is one of the prettiest desert birds. It is one of the few species that return to the same nest year after year.

Galahs are interesting birds, in that the number of eggs they lay depends on climate. If the desert is in a drought, they don't lay any. However, during more tolerable years, the galah may lay as many as five eggs.

Desert Mammals

There are several species of mammals in the desert. They range in size from a few inches to several feet in length. Like other desert wildlife, mammals have to find ways to stay cool and drink plenty of water. Many desert mammals are burrowers.

They dig holes in the ground and stay there during hot days. They return to the surface at night to feed. Hamsters, rats and their relatives are all burrowers. Not only do the burrows keep the animals cool, they are also a great place to store food.

Of course, not all animals have holes in the ground. Kangaroo and spiny anteater both live in the Australian desert region. Spiny anteaters are unusual mammals because they lay eggs.

The desert is also full of wild horses, foxes and jackals, which are part of canine family. And we can't forget the cats. Lions are found all over the deserts of southern Africa. They get their water from blood of their prey.

Camels: The Cars of the Desert

Camels could be included in the mammal section. Camels are the cars of the desert. Without them, people would have great difficulty crossing the hot terrain. There are two types of camels: Bactrian and dromedary. The main difference between the two is the number of humps. The Dromedaries have one hump and Bactrian have two. Both kinds are used by people, but only Bactrian's are found in the wild.

The camels are great for transportation because they use very little water. Camels can withstand very high temperatures without sweating. They also store fat in their humps for food. If a Bactrian camel travels a long distance without eating, its hump will actually get smaller.

iii. Decomposers

Due to poor vegetation the amount of dead organic matter is very less. As a result the decomposers are very few. The common decomposers are some bacteria and fungi, most of which are thermophile.

3.3.3 Aquatic Ecosystems

Freshwater Ecosystems

Freshwater occupies only a relatively small portion of the earth's surface, when compared with marine or terrestrial ecosystems as shown in the figure. But these ecosystems are very important from the point of view of humans, since human beings thrived wherever there was abundant freshwater, like the basins of the rivers. The Indus valley civilization, the Sumerian civilization, the Nile valley civilization all flourished in the river banks.

Freshwater habitats can be classified into:

- Standing-water or lentic, e.g., ponds, lakes, etc.

- Running water or lotic habitats, e.g., streams, rivers, etc.

Temperature in freshwater habitats does not show much variation, though it is often a limiting factor. The turbidity of water is due to the type and amount of suspended materials such as silt, clay particles, etc.

This is also an important limiting factor because it influences the ability of light to penetrate to a certain level in the water body. The current action, especially in streams have a very important role in the distribution of organisms. The dissolved gases such as O_2 and CO_2 are also limiting factors in any aquatic ecosystem.

In a pond or a lake, there are certain zones which have distinct features, mainly based on the depth of the water body and the light penetration. They are:

- Littoral zone, which is shallow.

- Limnetic zone, which has open water upto the zone of effective light penetration.

- Profundal zone, the bottom and deep water area, where there is a paucity of light.

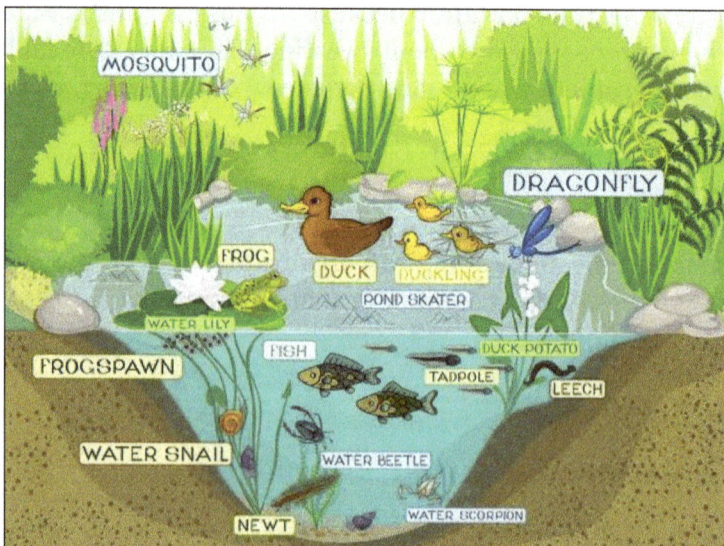

A pond ecosystem.

Lentic communities: Various organisms are found distributed in the various zones. In the littoral zone, the rooted plants, floating plants, emergent plants which are rooted plants which emerge out of the water level, submergement plants and phytoplanktons such as algae are present. The consumers are animals in which a vertical rather than horizontal zonation is evident.

In limnetic zones, the producers are mainly phytoplanktonic algae. In temperate lakes, phytoplanktonic population often shows a marked seasonal variation often leading to uncontrolled algal growth or blooms. The consumers of the limnetic zones are the zooplanktons, some insects and fishes.

In the profundal zone, the organisms mainly depend for their food on the littoral and limnetic zone. The region is rich in nutrients that are carried by currents and swimming animals to other zones.

Lotic community: In this ecosystem water current is a major limiting factor. The velocity of the current varies greatly in different parts of the stream or river. As streams flow for greater distances, the land-water interchange is often perceptible, with land forms extending into streams and the steams extending into land and other lentic habitats. As a result of the 'flowing' nature of the streams and rivers, oxygen tension is uniform and there is little or no thermal stratification.

Most of the plants are attached firmly to substrates in order to avoid getting carried away. The animals are either powerful swimmers or they get attached to the substrates by means of certain special structures such as hooks and suckers.

Marine (Ocean) Ecology

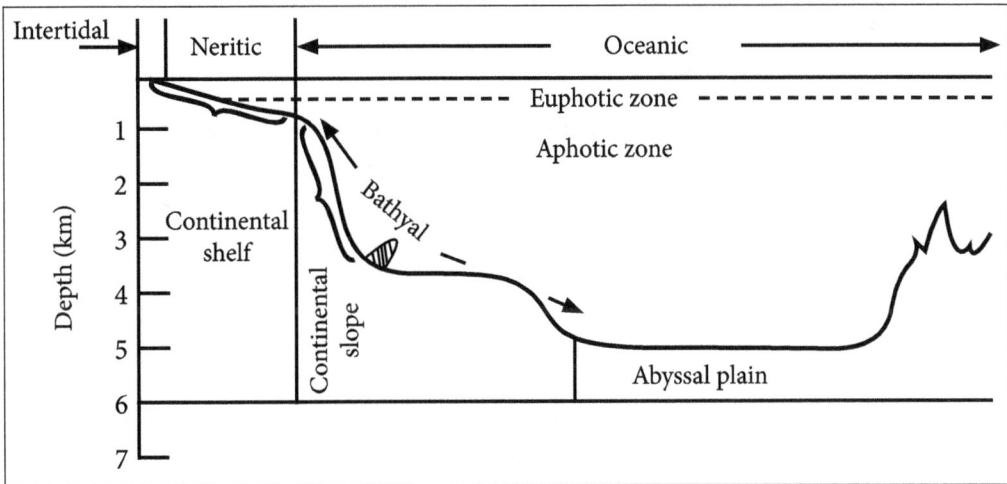

Zonation of sea.

The sea is very big, covering nearly 70% of all earth's surface. All the oceans are continuous. The sea is in continuous circulation because of the wind stress set up by air temperature differences between the pole and equator. The sea water is salty with an average of 35 parts per 1000 parts of water.

Zonation in the sea: The shallow water zone is termed the neritic zone. This is present in the continental shelf. The zone between the high and low tides is the intertidal zone. The continental shelf extending upto some distance offshore drops off steeply and is termed the continental slope.

The region beyond the continental shelf is termed the oceanic region, which has the continental slopes and beyond which is the abyssal plains. Based on the light penetration, the zones are termed euphotic and aphotic regions.

The communities: The marine organisms are very diverse and coelenterates, sponges, annelids, echinoderms, crustaceans and fish are dominant in marine waters. Algae are the most important phytoplanktons. The zooplanktons feed on the phytoplanktons. They include the protozoans, crustaceans, tiny jellyfish, free-floating polychaete worms, etc. They all remain as planktons in their entire life cycle and are termed as halo-planktons.

Meroplanktons comprise those animals, the larval forms of which are associated with the planktons. The benthos include a variety of organisms in the inshore region. They are a variety of crabs, amphipods, oysters, mussels, corals etc. The nekton and neuston include fishes, whales, seals, turtles, etc.

Estuarine Ecology

River mouths, coastal bays, tidal marshes and any other semi-enclosed coastal body of water which is influenced by tidal action and in which sea water is constantly in interaction with freshwater are called estuaries. Hence they are 'transitional ecotones' between freshwater and marine habitats.

Drowned river valleys, ford-type estuaries, bar-built estuaries, estuaries created by tectonic processes and river delta estuaries are some of the types of estuaries. Various estuaries have various amounts of salinity value, some being homogeneous, while others not. Estuaries are highly productive because of their 'being trapped' in between the sea and the freshwater ecosystem.

The shallow water production rate exceeds the rate of community respiration. The sea weeds, algal mats, sea grass beds, etc. which are profuse in growth are the primary producers. The estuaries also export energy and nutrients to deeper parts of the estuaries and to the sea. In the deeper parts, the nutrients are used up by the consumers. These are sedimentary in nature.

The plankton and nekton often move between both the shallow water and the deeper parts of the estuaries. They have certain periodicities which may be daily, tidal or periodic in nature.

3.4   Concept of Biodiversity

Biodiversity or biological diversity is the diversity of life. There are a number of definitions and measures of biodiversity. The term Biological Diversity was coined by Thomas Lovejoy in 1980, while the word Biodiversity itself was coined by W.G. Rosen in 1985.

Biodiversity is the sum of all the different species of animals, plants, fungi and microbial organisms living on Earth and the variety of habitats in which they live. Scientists estimate that upwards of 10 million and some suggest more than 100 million different species inhabit Earth. Each species is adapted to its unique niche in the environment, from the peaks of mountains to the depths of deep sea hydrothermal vents and from polar ice caps to tropical rain forests.

Biodiversity underlies everything from food production to medical research. Humans the world over use at least 40,000 species of plants and animals on a daily basis. Many people around the world still depend on wild species for some or all of their food, shelter and clothing. All of our domesticated plants and animals came from wild-living ancestral species.

Healthy ecosystems are very important to biodiversity. They regulate many of the chemical and climatic systems that makes the clean air and water and plentiful oxygen available. Forests, for example, regulate the amount of carbon dioxide in the air, produce oxygen and control rainfall and soil erosion. Ecosystems, in turn, depend on the continued health and vitality of the individual organisms that compose them. Removing just one species from an ecosystem can prevent the ecosystem from operating optimally.

Scientists have discovered and named only 1.75 million species less than 20% of those estimated to exist. And of those identified, only a fraction has been examined for potential medicinal, agricultural or industrial value. Much of the Earth's great biodiversity is rapidly disappearing, even before we know what is missing.

Most biologists agree that life on Earth is now faced with the most severe extinction episode since the event that drove the dinosaurs to extinction 65 million years ago. Species of plants, animals, fungi and microscopic organisms such as bacteria are being lost at alarming rates, in fact, that biologists estimate that three species go extinct every hour. Scientists around the world are cataloging and studying global biodiversity in hopes that they might better understand it or at least slow the rate of loss.

Classification of Biodiversity

Biodiversity is generally classified into three types:

- Genetic diversity.

- Species diversity.

- Community (or) Ecosystem diversity.

1. Genetic Diversity

A species with different genetic characteristics are known as subspecies or genera.

Genetic Diversity is Diversity within Species

Within individual species, there are number of varieties which are slightly different from one another. These differences are due to the differences in the combination of gases. Genes are basic units of hereditary information transmitted from one generation to other.

For Example:

i. Rice Varieties

All rice varieties belong to species "Organization". But there are thousands of rice varieties, which show variation at genetic level differ in their size, shape, color and nutrient content.

ii. Teak Wood Varieties

There are number of teak wood varieties available.

Example: Indian teak, Burma teak, Malaysian teak, etc

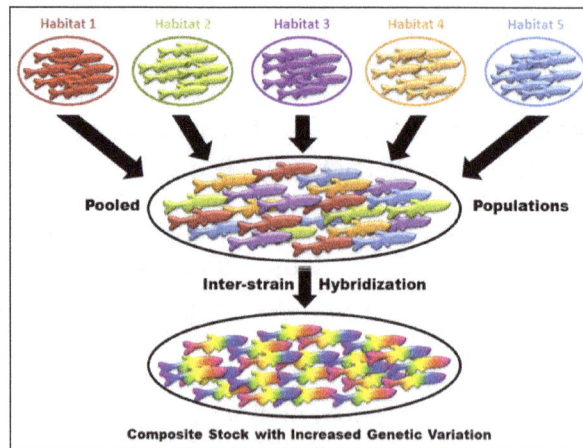

Genetic diversity.

2. Species Diversity

A discrete group of organisms of the same kind is known as species. Species diversity is the diversity between differed species. The sum of varieties of all the living organism at the species level is known as species diversity.

Biotic component is composed of a large number of species of plants, animals and microorganisms which interact with each other and with the abiotic component of the environment.

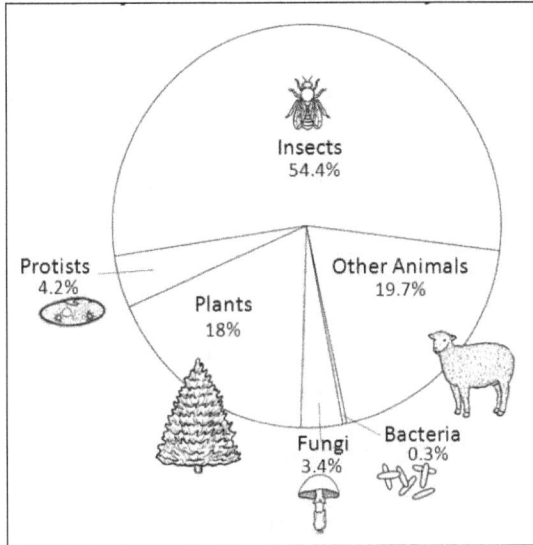

Species diversity Graph.

Example:

- Total number of living species in the earth are more than 20 million. But, of which only about 1.5 million living organisms are found and given scientific names.

- Plant species: Apple, mango, grapes, wheat, rice, etc.

- Animal species: Lion, tiger, elephant, dear, etc.

Community Ecosystem Diversity

Ecosystem

It is a set of biotic components interacting with one another and with abiotic components.

The diversity at the ecological or habitat level is known as ecosystem diversity. A large region with various ecosystems can be considered as ecosystem diversity.

Example: River ecosystem.

The river which include the fish, aquatic insects, muscles and variety of plants that have adapted. Thus, the ecosystem diversity is a aggregate of different environment types in a region. It explain the interaction between living organisms and physical environment in an ecosystems.

3.4.1 Bio-geographical Classification of India

The Indian mainland stretches from 8° 4' to 37° 6' N latitude and from 68° 7' to 97° 25' E longitude.

India is the seventh largest country in the world and Asia's second largest nation with an area of 3,287,263 square km. It measures 3,214 km from North to South and 2,933 km from East to West. It has a land frontier of some 15,200 kms and a coastline of 7,516 km (Government of India, 1985).

It is bounded on the south-west by the Arabian sea and on the south-east by the Bay of Bengal. India's northern frontiers are with Xizang (Tibet) in the Peoples Republic of China, Nepal and Bhutan. In the north-west, India borders on Pakistan in the north-east, China and Burma and in the east, Burma.

On the north, north-east and north-west lie the Himalayan ranges. Kanyakumari constitutes the southern tip of Indian peninsula, where it gets narrower and narrower, loses itself into the Indian Ocean. For administrative purposes India is divided into 28 states and 7 union territories.

Physically the massive country is divided into four relatively well defined regions:

- The Himalayan region,

- The Gangetic river plains or Indo-Gangetic plains,

- The Southern (Deccan) plateau,

- The Islands of Lakshwadeep, Andaman and Nicobar.

Bio-geographically, the country has 10 different biogeographic zones as shown in the figure. Biogeographic zones in India are:

- Trans-Himalayan Ladakh mountains,

- Himalayan region: The north-west, west, cental and eastern,

- The desert zone: Thar, Kutch and Ladakh,

- Semi-Arid: Central India, Gujarat — Rajawara,

- Western Ghats: Malabar coast, Western ghats,

- Deccan Peninsula: Central highlands, Chhota Nagpur, Eastern highlands, Central plateau, Deccan south,

- Gangetic plains: Upper and lower Gangetic plains,

- North-East India: Brahmputra valley, North-East hills,

- Islands: Andaman and Nicobar, Lakshadweep,

- Coast: West and East.

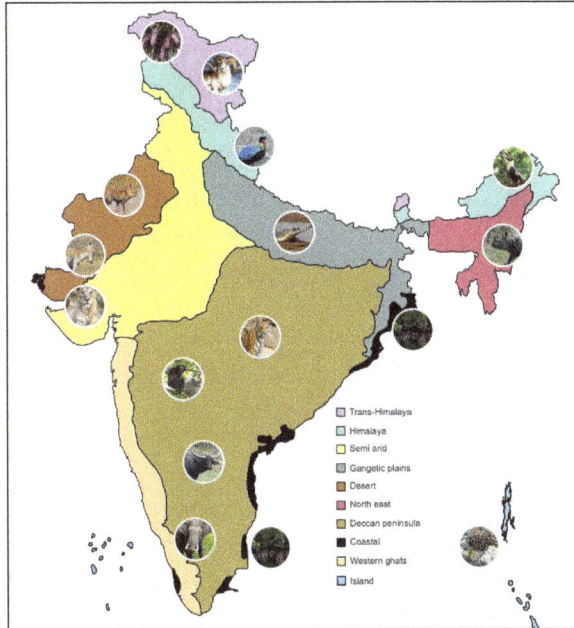

India's major Biogeographical zone.

Each of these ten biogeographic zones has characteristic biota and broadly represents similar climatic conditions and constitute the habitat for diverse species of flora and fauna. These biogeographic zones are discussed below:

1. The Trans-Himalayan Region

This area is very cold and arid. The only vegetation is a sparse alpine steppe. Extensive areas consist of bare rock and glaciers. The faunal groups best represented here are wild sheep and goats (chief ancestral stock), ibex, snow leopard, marbled cat, marmots and black-necked crane.

2. Himalayan Zone

This zone extends from north-west region of Kashmir to the east up to NEFA. There are three zones of vegetation corresponding to three climatic belts. These are:

- The Sub-montane or Lower Region (5000-6000.11): It has vegetation dominated by trees of Shorea rubusta, Dalbergia sissoo, Acacia catechu, Cedrala Corona, etc.

- The Montane or Temperate Zone (5500-12000 ft): The vegetation in this zone is dominated by Pinus excelsa, Cedrus deodara, Quercus lamellas, Cedrela, Eugenia, etc.

- The Alpine Zone (above 12000 ft): The vegetation include Betula utilis, Junipers and Rhododendion. At about 15000 ft plant growth is nil.

3. The Desert Zone

This zone comprises of Kutch, Thar and Ladakh. The Kutch and Thar region have very hot and dry summer and cold winter. Rainfall is less than 700 mm. The Ladakh region is a cold desert region. The natural vegetation consists of tropical thorn forests and tropical deciduous forests, sandy deserts with seasonal-salt marshes and mangroves are found in the main estuary. Typical shrubs are phog growing on sand dunes. Sewan grass covers extensive areas called palindrome.

The desert possess most of the major insect species, 43 reptile species and moderate birds endemism are found here. No niche of the Thar is devoid of birds. The black buck was once the dominant mammal of the desert region now confined only to certain pockets.

The Gazelle is the only species of the Indian antelope of which the femles have horns Nilgai the largest antelope of India and the wild ass, a distinct subspecies, is now confined to the Rann of Kutch which is also the only breeding site in the Indian subcontinent for the flamingo's. Other species like desert fox, great Indian bustard, and chinkara and desert cat are also found.

4. The Semi-Arid Region

The semi-arid region in the west of India includes the arid desert areas of Thar and Rajasthan extending to the Gulf of Kutch and Cambay and the whole Kathiawar peninsula. The natural vegetation consists of tropical thorn forests and tropical dry deciduous forest, moisture forests (extreme north) and mangroves.

The sandy plains have a few scattered trees of Acacia and Prosopis. The gravelly plains have Calotropis, Gymnosporia, etc. The rocky habitats are covered by bushes of Euphorbia while species of Salvadora and Tamarix occur mainly near saline depressions.

5. The Western Ghats

They cover only 5% of India's land surface but are home to more than about 4,000 of the country's plant species of which 1800 are endemic.

The monsoon forests occur both on the western margins of the ghats and on the eastern side where there is less rainfall. This zone displays diversity of forests from evergreen to dry deciduous.

The tropical evergreen forests are very luxuriant and multistoreyed having tall trees such as Tectona sandis, cedrela toona, etc. and many species of Bamboos In Nilgiri hills, trees such as Michell, nilagirilla and gordonia obtuse are most common. The Nilgiri langur, Lion tailed macaque, Nilgiri tahr, Malabar grey hornbill and most amphibian species are endemic to the Western Ghats.

6. The Deccan Peninsula

The Deccan Peninsula is a large area of raised land covering about 43% of India's total land surface. It is bound by the Satpura range on north. Western Ghats on west and Eastern Chats on the east. The elevation of the plateau varies from 900 mts., in the west to 300 mu. in the east.

There are four major rivers that support the wetlands of this region which have fertile black and red soil. Large parts are covered by tropical forests. Tropical dry deciduous forests occur in the northern, central and southern part of the plateau.

The eastern part of the plateau in Andhra Pradesh, Madhya Pradesh and Orissa has moist deciduous forests. Fauna like tiger, sloth bear, wild boar, gaur, sambar and chital are found throughout the zone along with small relict populations of wild buffaloes, elephants and barasingha.

7. The Gangetic Plains

The Gangetic plain is one of India's most fertile regions. The soil of this region is formed by the alluvial deposits of the Ganges and its tributaries. The Gangetic plains stretching from eastern Rajasthan through Uttar Pradesh to Bihar and West Bengal are mostly under agriculture.

The large forest area is under tropical dry deciduous forest and the southeastern end of the Gangetic plain merges with the littoral and mangroves regions of the Sunderbans. The fauna includes elephants, black gazelle, rhinoceros, Bengal florican, crocodile, freshwater turtle and a dense waterfowl community.

8. The North-East

Biological resources are rich in this zone. The tropical vegetation of north-east India is rich in evergreen and semi evergreen rain forest, moist deciduous monsoon forests, swamps and grasslands. Mammalian fauna includes 390 species of which 63% are found in Assam. The area is rich in smaller carnivores. The country's highest population of elephants are found here.

9. The Indian Islands

It is a group of 325 islands. Andaman to the north and Nicobar to the south. The two are separated by about 160 kms. By the Ten Degree Channel of the sea. The rainfall is heavy, with both North-east and South-west monsoons. At present, 21 of the 325 islands in the Andaman and Nicobar Island are inhabited. Many unique plants and animals are found here. About 2,200 species of higher plants are found here of which 200 are endemic.

The Andaman and Nicobar Islands have tropical evergreen forests and tropical semi evergreen forests as well as moist deciduous forests, littoral and mangrove forests. 112 endemic species of avifauna, the Andaman water monitor, giant robber crab, 4 species of turtles, wild boar, Andaman day gecko and the harmless Andaman water snake are found only in these islands. The Narcondam hornbill found only in Narcondam is a large forest bird with a big beak. Coral reefs are stretched over an area of 11,000 sq. km. in the Andaman and 2,700 sq. km. in Nicobar.

10. The Coastal Region

The natural vegetation consists of mangroves. Animal species include dugong, dolphins, crocodiles and avifauna. There are 26 species of fresh water turtles and tortoises in India and 5 species of marine turtles, which inhabit and feed in coastal waters and lay their eggs on suitable beaches. Tortoise live and breed mainly on the land.

The highest tiger population is found in the Sunderbans along the coast adjoining the Bay of Bengal. Lakshadweep consists of 36 major islands-12 atolls. 3 reefs and 5 submerged coral banks-make up this group of islands more than three hundred kilometers to the west of the Kerala coast. The geographical area is 32 sq. km. and the usable land area is 26.32 sq. km.

The fauna consists mainly of four species of turtles, 36 species of crabs, 12 bivalves, 41 species of sponges including typical coral or namental fishes and dugongs. A total of 104 scleractinian corals belonging to 37 genera are reported.

3.5   Value of Biodiversity

Biosphere is a life supporting system to the human beings. It is the combination of different organisms. Each organisms in the biosphere has its own significance. Biodiversity a vital for healthy biosphere. Biodiversity is must for the stability and proper functioning of the biosphere. We get benefits from other organisms in number of ways. Sometimes we realize the real value of the organism only after it is lost in this earth.

The values of biodiversity have been classified into 6 types.

These are also known as commodity values or use values and invalue products harvested by the people. These can be easily estimated by observing the activities of representative group of people, by monitoring collection points for normal products and by examining export/import statistics. These are further classified as:

1. Consumptive use Value

This value is assigned where the biodiversity product can be harvested and consumed directly. For example, fuel, food, drugs, fibre etc. These goods are consumed locally and do not figure in national and international market.

i. Food

A large number of wild plants are consumed by human beings as food. Nearly 80 to 90% of our food crops have been domesticated only from the tropical wild plants. A large number of wild animals are also consumed as food.

Example:

- In Himalayan region.

- Insects: Spider and wild herbivores are consumed by many tribal and non-tribal Communities in India.

ii. Drugs

Around 70% of modern medicines are derived from plant and plant extracts. 20,000 plant species are believed to be used medicinally, particularly in the traditional system of Ayurveda and Sidha.

Example:

- Germany alone use more than 2,500 Species of plants for medicinal purposes in Homeopathy and other systems of medicines.

- India uses 3000 species of plants in Ayurveda, Homeopathy and System of medicines.

iii. Fuel

Fire woods are directly consumed by villagers or tribal. The fossil fuels like coal, petroleum and natural gas are also the products of fossilized biodiversity.

2. Productive Use Values

These are commercially usable values where the product is marketed and sold in commercial markets, both national and international. These may include products like tusks of elephants, musk from musk deer, silk from silk-worm, wool from sheep, fir of many animals, lac from lac-insects, etc Many industries are dependent on these values.

Common are the paper and pulp industry, plywood industry, silk industry, textile industry, ivory-works industry, leather industry, pearl industry, etc.

Animal Product	Animal
Silk	Silk-worm
Wool	Sheep
Leather	All animals
Food	Fish and animals

Table: Plant products for various industries:

Plant Product	Industry
Wood	Paper and pulp industry, plywood industry
Cotton	Textile industry
Fruits, vegetables	Food industry
Leather	Leather industry

- Rice accounts for 22% of the cropped area and cereals accounts for 39% of the cropped area.

- Oil seed production also helped in saving large amount of foreign exchange spend on importing edible oils.

3. Social and Cultural Values

Biodiversity brings us many social and cultural benefits. It adds to the quality of life, providing some of the most beautiful and appealing aspects of our existence.

These values are associated with social life, customs, religion and pyscho-spiritual aspects of people. Many of the plants like Tulsi, Peepal, Mango, Lotus, Bael, etc, are considered as holy and sacred.

Their leaves, fruits or flowers are used in worship or the plant itself is worshipped. Many animals like Cow, Snake, Bull, Peacock, Owl, etc., also hold special social importance. Thus, biodiversity has distinct social values attached with different societies.

For Example:

1. Holy Plants

Many plants are considered as the holy plants in our country.

Example: lotus.

The leaves, fruits of these plants are used in worship.

2. Holy Animals

Many animals are also considered as holy animals in our country.

Example: Cow, Snake, Bull, Peacock.

3. Ethical Value

It involves values related to moral justification for conservation of biodiversity. It is based on the concept of Live and Let Live and involves ethical issues like all life must be preserved. These values are thus based on the belief that all species have a moral right to exist, independent of our need for them. These values are deep rooted within our society, religion and culture, but those who look on cost benefit analysis, they overlook these values.

4. Aesthetic Value

The beautiful nature of plants and animals insist us to protect the biodiversity. The most important aesthetic value of biodiversity is ecotourism.

Example:

- Eco-tourism: People from far place spend a lot of time and money to visit the beautiful areas, where they can enjoy the aesthetic value of biodiversity. This type of tourism is known as ecotourism.

- The pleasant music of wild birds.

- Color of butterfly, color of flowers, and color of peacocks are very important for their aesthetic value.

5. Option Values

The option values are the potential of biodiversity that are presently unknown and need to be known. The optimal values of biodiversity suggests that any species may be proved to be a valuable species after some day.

Example:

- The growing biotechnology field is searching a species for causing the disease of cancer and AID's.

- Medicinal plants and herbs play a very important role in our Indian economic growth.

3.5.1 Biodiversity at Global, National and Local Levels

There are at present 1.8 million species known and documented by scientists in the world. Although, scientists have estimated that the number of species of plants and animals on earth could vary from 1.5 to 20 billion. Thus the majority of species are yet to be discovered.

Biodiversity is the measure of variety of earth's animal, plant and microbial species of genetic differences within species and of the ecosystems that support the species. Out of an estimated 30 million species on earth, only one-sixth has been identified and authenticated in the past 200 years.

An estimated biodiversity covers 400,000 higher plants. Most of the world's bio-rich nations are in the South, which are developing nations. In contrast, the majority of the countries capable of exploiting biodiversity are Northern nations, in the economically developed world.

These nations however have low levels of biodiversity. Thus the developed world has come to support the concept that biodiversity must be considered to be a 'global resource'. Although, if biodiversity should form a 'common property resource' to be shared by all nations, there is no reason to exclude oil or uranium or even intellectual and technological expertise as global assets.

India's sovereignty over its biological diversity cannot be compromised without a revolutionary change in world thinking about sharing of all types of natural resources. The biodiversity of 89 countries with diversities higher than India are located in South America such as Brazil and South East Asian countries such as Malaysia and Indonesia.

The species found in these countries, however, are different from our own. This makes it imperative to preserve our own biodiversity as a major economic resource. While few of the other 'mega-diversity nations' have developed the technology to exploit their species for biotechnology and genetic engineering, India is capable of doing so.

Throughout the world, the value of biologically rich natural areas is now being increasingly appreciated as being of unimaginable value. International agreements such as the World Heritage Convention attempt to protect and support such areas. India is a signatory to the convention and has included several protected Areas as World Heritage sites.

These include Manas on the border between Bhutan and India, Kaziranga in Assam, Bharatpur in U.P., Nandadevi in Himalayas and the Sunderbanns in the Ganges delta in West Bengal. India has also signed the Convention in the Trade of Endangered Species (CITES) which is intended to reduce the utilization of endangered plants and animals by controlling trade in their products and in the pet trade.

Biologically, tropical rain forests are the centres of the world much of the earth's contemporary flora and fauna originated in the humid tropics. For millions of years, tropical rain forests have been factories of evolutionary diversity from which plants and animals, capable of adapting to more difficult environments, have gone forth to populate the subtropical and temperate regions.

It is essential to maintain areas of tropical rain forest large enough for this evolution to continue. The tropical forests are regarded as the richest in biodiversity. The species diversity in tropics is high.

The reasons are as follows:

- Warm temperate and high humidity provide favourable conditions for many species.

- Tropical communities are more productive because these areas receive more solar energy.

- Over geographical times the tropics had a more stable climate. In tropics, therefore, local species continued to live there itself.

- Among plant rates of out crossing appear to be higher in tropics.

The biodiversity exists on earth in eight broad realms with 193 bio-geographical provinces. Each bio-geographical province is composed of the ecosystems, which are constituted by communities of living species existing in an ecological region.

The developing countries, located in subtropical/tropical belt are far richer in biodiversity than the industrial countries in the temperate region. The Vavilovian Centres of diversity of crops and domesticated animals are also located in the developing countries.

It is important to preserve the numerous varieties of plants and animals that belong to one species. Each variety within a species contains unique genes and the diversity of genes within a species increases its capacity to adapt to pollution disease and other changes in the environment.

3.5.2 India as a Mega-diversity Nation

India has a great diversity of natural ecosystems from the cold and high Himalayan ranges to the sea coasts, from the wet northeastern green forests to the dry northwestern arid deserts, different types of forests, wetlands, island and the oceans. India consists of fertile river plains and high plateaus and several major rivers including the Ganges, Brahamaputra and Indus.

The climate of India is determined by the southwest monsoon between June and October, the northeast monsoon between October and November and dry winds from the north between December and February. From March to May the climate is dry and hot.

India, shows a great diversity in climate, topography and geology and hence the country is very rich in terms of biological diversity. India's biological diversity is one of the most significant in the world since India has only 2% of the total landmass of the world containing about 6% of the world known wildlife.

Based on reports of MOEF (Ministry of Forests and Environment), India at present has 89, 317 species of fauna and 45,364 species of flora about 7.3% of World's fauna and 10.88% the world flora. Out of the total animal species, insects alone comprise 68.52% and chordates 5.7%. Among the large animals, 173 species of mammals, 101 of birds, 15 of reptiles, 3 of amphibians and 2 of fishes are considered endangered.

India is also rich in Agro-biodiversity and ranks seventh in terms of contribution to world agriculture. There are 167 crop species and wild relatives. It is considered to be the centre of origin of 30,000-50,000 varieties of rice, pigeon-pea, turmeric, mango, ginger, sugarcane, gooseberries, etc.

India is also rich in marine biodiversity. It has a coastline of 7516.5 km., with exclusive economic zone of 200 million sq. km., supporting the ecosystems such as mangroves, estivaries, lagoons and coral reefs. The marine biodiversity includes:

- About 1600 species of zooplanktons.

- The benthic fauna consists of polychaeta (62%), crustacean (20%) and molluscs (18%).

- Over 30 species of marine algae, species of seagrass and over 45 species of mangrove plants.

- Over 342 species of corals and about 50% of the world's reef building corals are found in India.

3.5.3 Hotspots of Biodiversity

The hot spots are the geographic areas which possess the high endemic species.

At the global level, these are the areas of high conservation priority. If these species lost, they can neither be replaced nor regenerated.

Criteria for Recognizing Hot Spots

- The hot spots should have a significant percentage of specialized species.

- The richness of the endemic species is the primary criterion for recognizing hot spots.

- The site is under threat.

- It should contain important gene pools of plants of potentially useful plants.

Reasons for Rich Biodiversity in the Tropics

The following are the reasons for the rich biodiversity in the tropics :

- Warm temperatures and high humidity in the tropical areas provide favorable conditions.

- The tropics have a more stable climate.

- Among plants, rate of outcrossing appear to be higher in tropics.

- No single species can dominate and thus there is an opportunity for many species to coexist.

Hot Spots of Biodiversity in India

- Eastern Himalayas - Indo-Burma Region.

- Western Ghats - Sri Lanka Region.

Eastern Himalayas

Geographically this area comprises of Nepal, Bhutan and neighboring states of Northern India. There are 35,000 plant species found in the Himalayas, of which 30% are endemic.

The Eastern Himalayas are also rich in wild plants of economic value.

Examples: Rice, Banana, Juice and Sugarcane.

The taxol yielding plant is also sparsely distributed in the region.

- 63% mammals are from Eastern Himalayas.

- 60% of the Indian Birds are from North West.

- Huge wealth of fungi, insects, mammals, birds have been found in this region.

Western Ghats

The area comprises Maharashtra and Kerala. Nearly 1500 endemic, dicotyledons plant species are found from Western ghats. 62% amphibians and 50% lizards are endemic in Western Ghats. It is reported that only 6.8% of the original forests are existing today while the rest has been deforested or degraded.

Some common plants: Japonica, Rhododendron and Hypericum.

Some common animals: Blue bird, Lizard, hawk.

3.6   Threats to Biodiversity

Extinction is the Complete Elimination of Wild Species

It is Natural but slow process due to unplanned activities of man, the rate of decline of wild life has been particularly rapid in the last one hundred years. There are a number of causes which are known to cause the extinction of wildlife.

Wetlands have been drained to increase agricultural land. These changes have economic implications in the longer term. The current destruction of the remaining large areas of wilderness habitats, especially the super diverse tropical forests and coral reefs, is the most important threat worldwide to biodiversity.

Scientists have estimated that the human activities are likely to eliminate approximately 10 million species by the year 2050. There are around 1.8 million species of plants and animals, both large and microscopic, known to science in the world at present.

The number of species however is likely to be greater by a factor of at least 10. The plants and insects as well as other forms of life is not known to science are continually being identified in the worlds 'hot spots' of diversity.

Unfortunately at the present rate of extinction, about 25% of the worlds species will undergo extinction fairly rapidly. This may occur at the rate of 10 to 20 thousand species per year, a thousand to ten thousand times faster than the expected natural rate.

Much of this mega extinction spasm is related to human population growth, industrialization and changes in land use patterns. A major part of these extinctions will occur in 'bio-rich' areas such as tropical forests, wetlands and coral reefs. The loss of wild habitats due to rapid human population growth and the short term economic development are one of the major contributors to the rapid global destruction of bio-diversity.

Habitat Loss

It is the most serious threat to wildlife. It is due to Environmental pollution. Habitat loss also results from man's introduction of species from one area into another, disturbing the balance in existing communities. Loss of species occurs due to destruction of natural ecosystems, either for conversion to agriculture or industry or by over extraction of their resources or through pollution of air, water and soil.

Effects of Habitat Loss on Biodiversity

Habitat loss is a process of environmental change in which a natural habitat is rendered functionally unable to support the species present. It may be natural or unnatural and may be caused by habitat fragmentation, geological processes, climate change or human activities such as the introduction of invasive species or ecosystem nutrient depletion. In the process of habitat destruction or ganisms that previously used the site are displaced or destroyed, reducing biodiversity.

Human destruction of habitats has accelerated greatly in the latter half of twentieth century. Natural habitats are often destroyed through human activity for the purpose of harvesting natural resources for industry production and urbanization. Clearing habitats for agriculture, for example, is the principal cause of habitat destruction. Other important causes of habitat destruction include logging, mining and urban sprawl. Habitat destruction is currently ranked as the primary cause of species extinction worldwide.

Consider the exceptional biodiversity of Sumatra. It is home to one sub-species of orangutan, a species of critically endangered elephant and the Sumatran tiger, however half of Sumatra's forest is now gone. The neighboring island of Borneo, home to the other sub species of orangutan, has lost a similar area of forest and forest loss continues in protected areas.

The orangutan in Borneo is listed as endangered by the International Union for Conservation of Nature (IUCN), but it is simply the most visible thousands of species which does not survive on the disappearance of the forests of Borneo. The forests are being removed for their timber and to clear the space for plantations of palm oil, an oil used in Europe for many items including food products, cosmetics and bio-diesel.

Poaching of Wildlife

Poaching is another threat to wildlife. In ancient period hunters, Collectors and smugglers (traders) are the major threat to a number of species including endangered species. They collected furs, hides, horns, tusks and some live specimens, herbal products and smuggled to others for millions of dollars. The alarming point in this case is that for one animal they killed more than one. It is an illegal trade and internationally banned.

The cost of these animal parts are surprising. The cost of Bengal tiger coat is more than one lac dollars. South American ocelot cost more than 50,000 dollars, a single orchid cost more than 5000 dollars, horns of rhinoceros cost their weight in gold. These are some examples by which we can understand the situation of trading wildlife products, which is highly profit making for poachers.

Over collection and over exploitation are the main causes of disappearance of plants of scientific and medicinal value. The reduction of genetic diversity among the cultivated species drastically limit possibilities of creating new cultivar in the future, which could be disastrous for human race. It is advisable that do not purchase the parts and products made from wild animals specially endangered species.

Man-Wildlife Conflicts

Struggle for Existence

This is applicable for both, man and wild animal. Due to habitat loss animals come out of the forest and destroy the crops later on they become danger to human being. Villagers and affected people kill them. There are so many cases of conflict between man and wild life. In these cases forest department could not pacify resulting in lack of non-co-operation of wild life conservation from affected people.

Animals are prone to infection when they are under stress. Animals held in captivity are also more prone to diseases. The elephants and other wild animals suffer pain and turn violent when they tend to destroy the electric fenced crop field. It is noted that ill, weak and injured animals have tendency to attack man. Man and wild life conflicts also occur during human encroachment into forest area. There are number of case.

Depletion of Income

Income has depleted of rural population due to check on hunting, harvesting and human conflict is an increasing example. The elephants being cute and lovable are the cherish able animals. But if they are put next to our house they turn violent leading to conflict with human.

Biodiversity of an area influences every aspect of lives of the people who inhabit it. Their livelihood and their living space depends on the type of ecosystem. Even people living in urban areas are dependent on the ecological services provided by the wilderness in the PAs. It is linked with the service that nature provides us.

The quality of water we drink and use, the air we breathe, soil on which our food grows are all influenced by a wide variety of living organisms both plants and animals and the ecosystem of which each species is linked with in nature.

While it is well known that plant life removes the carbon dioxide and releases oxygen we breathe, it is less obvious that fungi, small soil invertebrates and even microbes are essential for plants to grow.

A natural forest maintains the water in the river after the monsoon or the absence of ants could destroy life on earth, are to be appreciated to understand how we are completely dependent on the living 'web of life' on earth. This includes mankind as well. Think about this and we cannot but want to protect out earth's unique biodiversity. We are highly dependent on these living resources.

3.6.1 Endangered and Endemic Species of India

According to International Union of Conservation of Nature and Natural Resources (IUCN), the species are classified into the following types:

- Extinct Species: A species is declared as extinct, when it is no longer found in the world.

Endangered species.

- Endangered Species: A species is declared as endangered, when its number has been reduced to a critical level. Unless it is under protection and conserved, it is in immediate danger of extinction.

- Vulnerable Species: A species is said to be vulnerable when its population is facing continuous decline due to habitat destruction of over exploitation. Such a species is still abundant.

- Rare Species: A species is said to be rare, when it is localized within restricted area (or) they are thinly scattered over a more extensive area. Such species are not endangered or vulnerable.

Endemic Species of India

Statistics

Category	Species Enlisted	Highly Endangered Species
Higher plants	15,000	135
Mammals	372	69
Birds	1,175	40
Reptiles & Amphibians	580	22
Fish	1,693	N.A

Equally disturbing and the matter of far more consequence is that we have till now not explored even 10% of the existing biodiversity and with such an alarming rate of extinction, we have even been destroying species, the potential of which is either not studied or less studied. In other words, the species are being destroyed with being even discovered or classified.

The causes for loss of species are complex and varied and prominent among these could be listed as follows:

- Modification, degradation and loss of habitats due to colonization and clearing of forest areas for settlement or agricultural expansion, commercial lodgings, large hydel schemes, fire, human and livestock pressure etc.

- Over exploitation, mainly for commercial (and often illegal) purposes like meat, fur, hides, body organs, medicinal etc.

- Accidental or deliberate introduction of exotic species which can threaten native flora and fauna directly by predation or by competition and also indirectly by altering the natural habitat or introducing diseases.

- Pollution (both air and water) stresses ecosystem, mismanagement of industrial and agriculture wastes threaten both terrestrial and aquatic ecosystem.

- Increase in the global surface temperature by 2° C to 6° C (global warming).

- The other possible reasons for loss of species could be improper use of agro-chemicals and pesticides, a rapidly growing human population, inequitable land distribution, economic and political policies and constraints.

3.7   Conservation of Biodiversity

Conservation is the protection, preservation, management or restoration of wildlife and natural resources such as forests and water. Through the conservation of biodiversity the survival of many species and habitats which are threatened due to human activities can be ensured.

Other reasons for conserving biodiversity include securing valuable natural resources for future generations and protecting the well-being of ecosystem functions.

In-Conservation

In- conservation involves protection of fauna and flora within its natural habitat, where the species normally occurs.

The natural habitats or ecosystems maintained under in-conservation are called as "protected areas".

Important In-Conservation

Biosphere reserves, national parks, wildlife sanctuaries, Gene sanctuary, etc.

Methods of In-Conservation

Around 4% of the total geographical area of the country is used for in-conservation. The following methods are presently used for in-conservation.

In- conservation	No. available
Biosphere reserves	7
National parks	80
Wild-life sanctuaries	420

1. Biosphere Reserves

Biosphere reserves cover large area, more than 5000 sq. km. It is used to protect species for long time.

Table: Biospheres Reserves in India:

S. No.	Name	Area of Biosphere (sq km)	Date of establishment	District	State
1	Agasthyamalai	1701.00	2001	-	Kerala
2	Achanakmar-Amarkantak	3835.51	2005	Anuppur, Dindori & Bilaspur	Madhya Pradesh & Chhattisgarh
3	Dibru-Saikhowa	765.00	1997	Dibrugarh and Tinsukia	Assam
4	Dehang-Debang	5111.5	1998	Siang & Debang Valley	Arunachal Pradesh
5	Gulf of Mannar	10,500.00	1989	Indian part of Gulf of Mannar	Tamil Nadu
6	Great Nicobar	885.00	1989	Southernmost Island of Andaman and Nicobar	Andaman and Nicobar
7	Manas	2837.00	1989	Part of Kokrajhar, Bongaigaon, Barpeta, Nalbari, Kamrup and Darang	Assam
8	Kanchanjunga	2619.92	2000	Kanchanjunga Hills	Sikkim
9	Nilgiri	5520.4	1986	Part of Wayanad, Bandipur and Nagarhole, Nilambur, Silent Valley and Siruvani Hills	TamilNadu, Kerala and Karnataka

10	Nanda Devi	5860	1988	Chamoli, Almora and Pithoragarh	Uttaranchal
11	Pachmarhi	4926.00	1999	Betul, Hoshang-abad and Chhind-wara	Madhya Pradesh
12	Nokerek	80.00	1988	Part of Garo Hills	Meghalaya
13	Sunderbans	9630.00	1989	Delta of Ganges and Brahmaputra	West Bengal
14	Similipal	4374.00	1994	Mayurbhanj	Orissa

Role of Biosphere Reserves

- It gives long-term survival of evolving ecosystem.

- It protects endangered species.

- It serves as site of recreation and tourism.

- It is also useful for educational and research purposes.

Restriction

No tourism and explosive activities are permitted in the biosphere reserves.

2. National Park

A National Park is an area dedicated for the conservation of wildlife along with its environment. It is usually a small reserve covering an area of about 100 to 500 sq.ft. Within the biosphere reserves, one or more national parks are also exists.

Role of a National Park

- It is used for enjoyment through tourism, without affecting the environment.

- It is used to protect, propagate and develop the wildlife.

Restrictions

- Grazing of domestic animals inside the national park is prohibited.

- All private rights and forestry activities are prohibited within a national park.

3. Wildlife Sanctuaries

A wildlife sanctuary is an area, which is reserved for the conservation of animals only. At present, there are 492 wildlife sanctuaries in our country.

Role of Wildlife Sanctuaries

- It allows the operations such as collection of forests products, harvesting of timer, private ownership rights and forestry operations provided.

- It does not affect the animals adversely.

- It protects animals only.

Restrictions

Killing, hunting, shooting or capturing of wildlife is prohibited except under the control of higher authority.

4. Gene Sanctuary

A gene sanctuary is an area, where the plants are conserved:

Example: In Northern India, two gene sanctuary are found available.

- One gene sanctuary for citrus.

- One gene sanctuary for pitcher plant.

Advantages of In-Conservation

- It is cheap and convenient method.

- The species gets adjusted to the natural disasters like drought, floods, and forest fires.

Disadvantages

- Maintenance of the habitats is not proper, due to shortage of staff and pollution.

- Large surface area of the earth is required to preserve the biodiversity.

Ex-Conservation

Ex-conservation involves protection of fauna and flora outside the natural habitats.

This type of conservation is mainly done for conservation of crop varieties and the wild relatives of crops.

Role of Ex-Conservation

- It identifies those species which are at more risk of extinction.

- It involves maintenance and breeding of the endangered plant and animal species under controlled conditions.

- It prefers the species, which are more important to man in near future among the endangered species.

Important Ex-Conservation

Botanical gardens, seed banks, microbial culture collections, tissue and cell cultures, museums.

Methods of Ex- Conservation

The following important gene bank (or) seed bank facilities are used in Ex-conservation:

i. National Bureau of Plant Genetic Resources (NBPGR):

It is located in New Delhi. It uses preservation techniques to preserve agricultural and horticultural crops.

Preservation Technique

It involves the preservation of seeds, pollen of some important agricultural and horticultural crops by using liquid nitrogen at a temperature as low as - 196°C. Varieties of rice, turnip, radish, onion, have been preserved successfully in liquid nitrogen for several years.

ii. National Bureau of Animal Genetic Resources (NBAGR):

It preserves the semen of domesticated bovine animals.

iii. National Facility for Plant Tissue Culture Repository (NFPTCR):

It develops the facility for conservation of varieties of crop plants or tree by tissue culture. This facility has been created within the NBPGR.

Advantages of Ex-Conservation

- Survival of endangered species is increasing due to special care and attention.

- In captive breeding, animals are assured food, water, shelter and also security and hence longer life Span.

- It is carried out in cases of the endangered species, which do not have any chances of survival, in the wild.

Disadvantages

- The animals cannot survive on natural environment.

- It is expensive method.

- The freedom of wildlife is lost.

Chapter 4

Environmental Pollution

4.1 Environmental Pollution and its Various Types

Our surrounding environment is becoming contaminated day by day, by the increased addition of domestic as well as industrial wastes to it. The water, air and land which are essential for living beings should be in their pure state.

Modern civilization, increasing population and growth of industries have altered the natural environment. The air we breathe, the water we drink and the places where we live are contaminated with toxic substances resulting in air, water and land pollution. These results in many health hazards.

Pollution may be defined as the excessive discharge of undesirable substances into the environment, adversely altering the natural quality of the environment and causing damage to humans, plants and animals.

Pollution may also be defined as unwanted or detrimental changes in a natural system. A pollutant is the undesirable foreign matter added to the environment.

Causes, Effects and Control Measures of Air Pollution

The presence of one or more contaminants like smoke, dust, mist and odor in the atmosphere which are injurious to human beings, plants and animals.

According to the World Health Organization (WHO), more than 1.1 billion people live in urban areas where outdoor air is unhealthy to breath. Some of the common air pollutants are as follows:

1. Carbon Monoxide (CO)

It is a colorless gas that is poisonous to air breathing animals. It is usually formed during the incomplete combustion of carbon containing fuels,

$$2C + O_2 \rightarrow 2CO$$

Human Source: Cigarette smoking, incomplete burning of fossil fuels. About 77% comes from motor vehicle exhaust.

Health Effects: Reacts with hemoglobin in red blood cells and reduces the ability to bring oxygen to body cells and tissues by the blood. Which can cause anemia and head-aches. At high levels it leads to coma, irreversible brain cell damage and death.

2. Nitrogen Dioxide (NO_2)

Nitrogen dioxide is a reddish-brown irritating gas that gives the photochemical smog. In the atmosphere it can be converted into nitric acid.

$$NO_2 + Moisture \rightarrow HNO_3$$

Human Source: Fossil fuel burning in motor vehicles (49%) and power industrial plants (49%).

Health Effects: Lungs irritation and damage.

Environmental effects: Acid deposition of HNO_3 can damage trees, soils and aquatic life in lakes, HNO_3 can corrode metals and eat away stone on buildings, statues and monuments. NO_2 can damage fabrics.

3. Sulfur Dioxide (SO_2)

Sulfur dioxide is a colorless and irritating gas. It is formed mostly from the combustion of sulfur containing fossil fuels such as coal and oil. In the atmosphere it can be converted to sulfuric acid which is a major component of acid deposition.

Human Sources: Coal burning in power plants (88%) and industrial processes (10%).

Health Effects: Breathing problems for healthy people.

Environmental effects: Reduce visibility.

Acid deposition of H_2SO_4 can damage trees, soils and aquatic life in lakes.

4. Hydrocarbons

Hydrocarbons, especially lower hydrocarbons gets accumulator due to the decay of vegetable matter.

Human source: Agriculture, decay of plants and burning of wet logs.

Health effects: Carcinogenic.

5. Ozone (O_3)

It is a highly reactive irritating gas with an unpleasant odor that is formed in the tropo-sphere. It is a major component of the photochemical smog.

Human Source: Chemical reaction with volatile organic compounds and nitrogen oxides.

Environmental Effects: Moderates the climate.

6. Chromium (Cr)

It is solid toxic metal, emitted into the atmosphere as particulate matter.

Human Source: Paint, Smelters, Chromium plating.

Health Effects: Perforation of nasal septum, chrome holes.

Control Measures

The atmosphere has several built-in self-cleaning processes such as dispersion, gravitational settling, flocculation, absorption, rain washout and so on, to cleanse the atmosphere.

In terms of a long range of air pollution, control of contaminants at their source is a more desirable and effective method through preventive or control technologies:

i. Source Control

Since we know the substances that causes air pollution, the first approach to its control will be through source reduction:

- Use only unleaded petrol.

- Use petroleum products and other fuels that have low sulfur and ash content.

- Reduces the number of private vehicles on the road by developing an effective public transport system.

- Ensure that houses, schools, restaurants and places where children play are not located on busy streets.

- Plant trees along busy streets because they remove particulates and carbon monoxide and absorb noise.

- Industries and waste disposal sites should be situated outside the city center preferably downwind of the city.

- Use catalytic converters to help control the emission of carbon monoxide and hydrocarbons.

ii. Control Measures in Industrial Centers

- The emission rates should be restricted to permissible levels by each and every industry.

- Continuous monitoring of the atmosphere for the pollutants should be carried out to know the emission levels.

- Incorporation of air pollution control equipment's in the design of the plant layout must be made mandatory.

Public Health Aspects

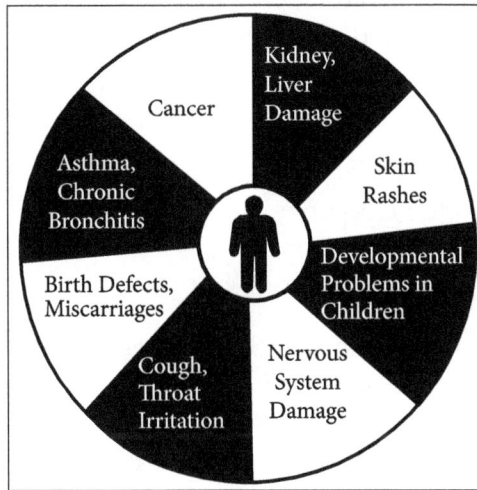

Effects on public health.

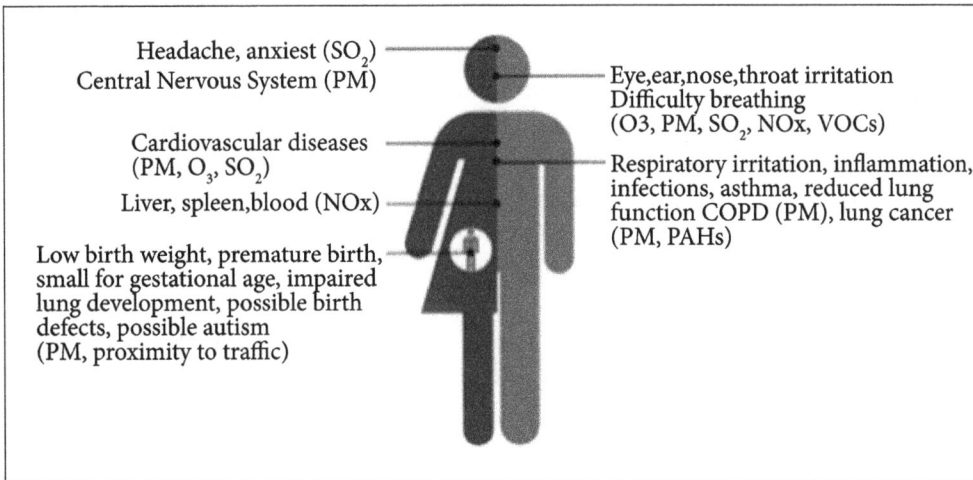

The Air Pollution and Respiratory Health Branch leads CDC's fight against the environmental related respiratory illnesses, including asthma and studies indoor and outdoor air pollution.

Air Pollution and Repository Health

- Asthma: It is a serious environmental health threat, but it can be controlled by taking the medication and by avoiding contact with the environmental "triggers" such as furry pets, dust mites, tobacco smoke, mold and certain chemicals.

- Mold: Exposure to damp and moldy environments may cause throat irritation, nasal stuffiness, eye irritation, coughing or wheezing or skin irritation.

- Carbon Monoxide Poisoning: Carbon monoxide (CO), which is an odorless, colorless gas that can cause sudden illness and death, is found in combustion fumes produced by generators, cars and trucks, lanterns, stoves, gas ranges, burning charcoal and wood and heating systems.

Climate and Health

Climate and Health Program works to prevent and adapt to the health impacts of the extreme weather and the other climate related issues.

Environmental Public Health Tracking

Environmental Public Health Tracking is the ongoing collection, the integration, analysis, interpretation and dissemination of data on environmental hazards, exposures to those hazards and health effects which may be related to the exposures. The goal of tracking is to provide information which can be used to plan, apply and evaluate actions to prevent and control environmentally related diseases.

Health Studies

The Health Studies investigates the human health effects of exposure to environmental hazards ranging from chemical pollutants to natural, technologic or terrorist disasters. The results are used to develop, implement and evaluate strategies which can help in preventing or reducing harmful exposures.

Radiation Studies

The Radiation Studies identifies the potentially harmful environmental exposures to ionizing radiation and associated toxicants, conducts the energy related health research and responds to protect public's health in the event of an emergency involving radiation or radioactive materials.

4.1.1 Water Pollution

Water pollution is the pollution caused due to any chemical, physical or biological change in water quality that has a harmful effects on living organisms or makes water unsuitable for desired uses.

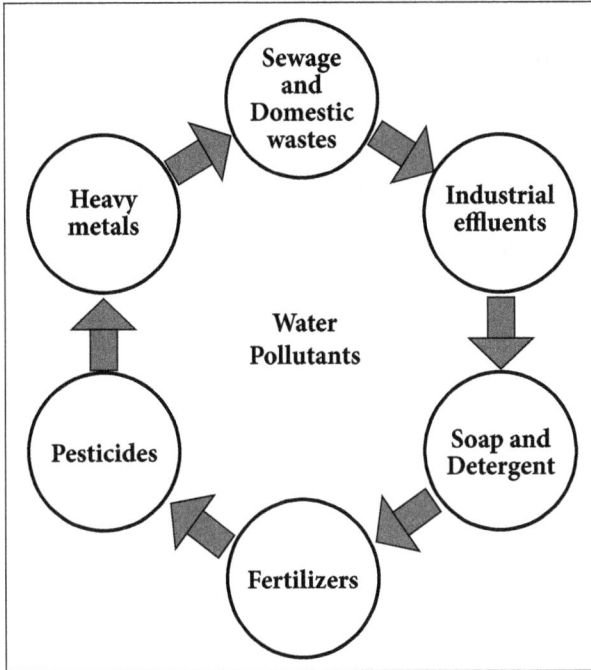

Water pollutants.

1. Infection Agents

Example: Bacteria, Viruses, Protozoa.

Human Sources: Human and animal wastes.

Effects: Variety of diseases.

2. Oxygen Demanding Wastes

Example: Organic wastes such as animal manure and plant debris that can be decomposed by aerobic bacteria.

This degradation consumes dissolved oxygen in water. DO is the amount of oxygen dissolved in a given quantity of water at a particular pressure and temperature.

The standard point of DO varies from 8 - 15 mg/lit.

Human Source: Sewage, animal feedlots, paper mills.

Effects: Large populations of bacteria decomposing these wastes can degrade water quality by depleting water of dissolved oxygen. This causes fish and other forms of oxygen consuming aquatic life to die.

3. Organic Chemicals

Example: Oil, Gasoline, Plastics, Pesticides.

Human Sources: Industrial effluents, surface runoff from farms, household cleansers.

Effects:

- Can threaten human health by causing nervous system damage and some cancers.
- Harm fish and wildlife.

4. Sediment

Examples: Soil, Silt, etc.

Human Sources: Land erosion.

Effects:

- Can cloud water and reduce photosynthesis.
- Disrupt aquatic food webs.
- Carry pesticides, bacteria and other harmful substances.
- Settle out and destroy feeding and spawning grounds of fish.

5. Radioactive Materials

Example: Radioactive isotopes of iodine, radon, uranium and chromium.

Human Sources: Nuclear power plants, mining and processing of uranium and other ores, nuclear weapons production and natural sources.

Effects: Genetic mutations, birth defects and certain cancers.

Control Measures of Water Pollution

- Administration of water pollution control should be in the hands of state or central government.
- Scientific techniques should be adopted for environmental control of catchment areas of rivers, ponds or streams.
- Industrial plants should be based on recycling operations as it helps to prevent the disposal of wastes into natural waters but also the extraction of products from waste.

- Plants, trees and forests control the pollution as they act as natural air conditioners.

- Trees are capable of reducing sulphur dioxide and nitric oxide pollutants and hence more trees should be planted.

- No type of waste (treated, partially treated or untreated) should be discharged into any natural water body. Industries should develop closed loop water supply schemes and domestic sewage must be used for irrigation.

- Qualified and experienced people must be consulted from time to time for effective control of water pollution.

- Public awareness must be initiated regarding adverse effects of water pollution using the media.

- Laws, standards and practices should be established to prevent water pollution and these laws should be modified from time to time based on current requirements and technological advancements.

- Basic and applied research in public health engineering should be encouraged.

Soil Pollution

Soil pollution is caused due to the addition of chemicals which reduce its productive capacity. In addition to the urban solid wastes, several hazardous chemicals are dumped into the soil. The toxic substances from these dumps leach out and percolate through the soil and contaminate the ground water. In agricultural operations, most of fertilizers and pesticides are used and the residual chemicals remain in the top layers of soil. Toxic insecticides kill the useful soil bacteria that are favorable for plant growth.

Soil acts like a filter in removing the impurities in water and waste waters. But toxic residual chemicals from the soil reach human beings through vegetables, fruits, etc. Industrial effluents and solid wastes without adequate treatment which get deposited on the land or in water bodies are major sources of soil pollution.

Various pollutants and their harmful effects on soil are as follows:

- Organic wastes enter the soil pores and decompose. Pathogenic bacteria present in these wastes spread infection. Diseases caused by Hookworms and helminthus are common in villages due to soil pollution.

- Compounds containing arsenic, mercury, chromium, nickel, lead, cadmium, zinc and iron are toxic to life. Fluorides also affect the plant growth.

- Excess use of sodium, magnesium, calcium, potassium, sulfur, zinc and iron in the form of fertilizers and pesticides inhibit plant growth and reduce crop yield. It is necessary to keep the dosage at an optimal level.

- Water logging and salinity increases the dissolved salt content in ground water and the soil. Some plants are too much sensitive to soil pH and salinity. High salinity may make the land unfit for cultivation.

Soil erosion also causes soil pollution as given below:

- Soil erosion is the removal of soil from its original place. Water, ice wind and other climatic agents promote soil erosion. Cutting of trees, construction and other human activities reduce the vegetation cover which protects the soil from sun, rain, wind and moving water and as a result the soil is eroded.

- Due to soil erosion essential minerals are removed resulting in the loss of fertility of soil. Soil erosion results in the increased sedimentation of rivers and lakes which has an impact on water quality and subsequently on the populations of aquatic organisms including fish.

- It is not advisable to construct any new building on soils such as clay and silt that have been recently carried forward and deposited by the waters of rivers and streams as the load bearing capacity of such a soil medium is very low. There are cases where the underground stratum is peat deposited by sewage canals that had existed before. Such sites should never be selected for the construction of buildings.

- In a natural ecosystem, essential nutrients and minerals cycle from the soil to living organisms and back to the soil when the organisms die. This naturals system is disrupted by agricultural operations. In dry climates much of the water used for irrigation evaporates, leaving behind high concentration of salts such as sodium chloride in the top soil. The accumulation of salts in the top soil is known as salinization.

- For the conservation of soil, various methods are used to reduce soil erosion, to prevent depletion of soil nutrients and to restore nutrients already lost by erosion, leaching and excessive crop harvesting.

Control Measures of Soil Pollution

1. Soil Erosion

Soil erosion can be controlled by variety of forestry and farm practices.

Example: Planting trees on barren slopes:

- Contour cultivation and strip cropping may be practiced instead of shifting cultivation.

- Terracing and building diversion channels may be undertaken.

Reducing the deforestation and substituting chemical manures by animal wastes also helps arrest soil erosion in long term.

2. Proper Dumping of Unwanted Materials

Excess wastes by man and animals pose a disposal problem. The open dumping is the most commonly practiced technique. Nowadays, controlled tipping is followed for solid waste disposal. The surface so obtained is used for housing or sports field.

3. Production of Natural Fertilizers

Bio-pesticides should be used in place of toxic chemical pesticides. The organic fertilizers should be used in place of synthesized chemical fertilizers.

Example: The organic wastes in animal dung may be used to prepare compost manure instead of throwing them wastefully and polluting the soil.

4. Proper Hygienic Condition

The people should be trained regarding sanitary habits.

Example: Lavatories should be equipped with quick and effective disposal methods.

5. Public Awareness

Informal and formal public awareness programs should be imparted to educate people on health hazards by environmental education.

Example: Mass media, Educational institutions and voluntary agencies can achieve this.

6. Recycling and Reuse of Wastes

To minimize soil pollution, the wastes such as paper, plastics, metals, glasses or ganics, petroleum products and industrial effluents etc. should be recycled and reused.

Example: The industrial wastes should be properly treated at source. Integrated waste treatment methods should be adopted.

7. Ban on Toxic Chemicals

The ban should be imposed on chemicals and pesticides like DDT, BHC, etc which are fatal to plants and animals. The nuclear explosions and improper disposal of radioactive wastes should be banned.

Marine Pollution

It occurs when harmful or potentially harmful, effects result from the entry into the ocean of chemicals, particles, industrial, agricultural and residential waste, noise or spread of invasive organisms. Most sources of marine pollution are land based.

The pollution often comes from non-point sources such as agricultural runoff, wind-blown debris and dust. Nutrient pollution, a form of water pollution, refers to contamination by excessive inputs of nutrients. It is a primary cause of eutrophication of surface waters, in which excess nutrients, usually nitrogen or phosphorus, stimulate algae growth.

Causes of Marine Pollution

There are various ways for how pollution enters the ocean:

1. Sewage

Sewage or polluting substances flow through sewage, rivers or drainage's directly into the ocean. This is often how minerals and substances from mining camps find their way into the ocean. The release of other chemical nutrients into the ocean's ecosystem leads to reductions in oxygen levels, the decay of plant life, a severe decline in the quality of the sea water itself. As a result, all levels of oceanic life, plants and animals, are highly affected.

2. Toxic Chemicals from Industries

Industrial and agricultural wastes are another most common form of wastes that are directly discharged into the oceans, resulting in ocean pollution. The dumping of toxic liquids in the ocean directly affects the marine life as they are considered hazardous and secondly, they raise the temperature of the ocean, known as thermal pollution, as the temperature of these liquids is quite high. Animals and plants that cannot survive at higher temperatures eventually perish.

3. Land Runoff

It is another source of pollution in the ocean. This occurs when water infiltrates the soil to its maximum extent and the excess water from rain, flooding or melting flows over the land and into the ocean. Often times, this water picks up man-made, harmful contaminants that pollute the ocean, including fertilizers, petroleum, pesticides and other forms of soil contaminants. The fertilizers and waste from land animals and humans can be a huge detriment to the ocean by creating dead zones.

4. Large Scale Oil Spills

Ship pollution is a huge source of ocean pollution, the most devastating effect of which is oil spills. Crude oil lasts for years in the sea and is extremely toxic to marine life, often suffocating marine animals to death once it entraps them. Crude oil is also extremely difficult to clean up, unfortunately meaning that when it is split, it is usually there to stay.

5. Ocean Mining

Ocean mining in the deep sea is yet another source of ocean pollution. Ocean mining sites drilling for silver, gold, copper, cobalt and zinc create sulfide deposits up to three and a half thousand meters down into the ocean.

While we are yet to gather scientific evidence to fully explain the harsh environmental impacts of deep sea mining, we do have a general idea that deep sea mining causes damage to the lowest levels of the ocean and increase the toxicity of the region. This permanent damage dealt also causes leaking, corrosion and oil spills that only drastically further hinder the ecosystem of the region.

6. Littering

Pollution from the atmosphere is, believe it or not, a huge source of ocean pollution. This occurs when objects that are far inland are blown by the wind over long distances and end up in the ocean. These objects can be anything from natural things like dust and sand, to man-made objects such as debris and trash. Most debris, particularly plastic debris, cannot decompose and remains suspended in the oceans current for years.

Animals can become snagged on the plastic or mistake it for food, slowly killing them over a long period of time. Animals who are most often the victims of plastic debris include turtles, dolphins, fish, sharks, crabs, sea birds and crocodiles.

In addition, the temperature of the ocean is highly affected by carbon dioxide and climate changes, which impacts primarily the ecosystems and fish communities that live in the ocean. In particular, the rising levels of CO_2 acidify the ocean in the form of acid rain. Even though the ocean can absorb carbon dioxide that originates from the atmosphere, the carbon dioxide levels are steadily increasing and the ocean's absorbing mechanisms, due to the rising of the ocean's temperatures, are unable to keep up with the pace.

Effects of Ocean Pollution

- Effect of Toxic Wastes on Marine Animals.
- Disruption to the Cycle of Coral Reefs.
- Depletes Oxygen Content in Water.
- Failure in the Reproductive System of Sea Animals.
- Effect on Food Chain.
- Affects Human Health.

Noise Pollution

The unwanted, unpleasant or disagreeable sound that causes discomfort for all living things.

Noise pollution.

The sound intensity is measured in decibel (dB), which is tenth part of the longest unit Decibel. One dB is equal to the faintest sound, a human ear can hear.

Types of Noise

It has been found that the environmental noise is being doubling for every 10 years. Generally noise is described as:

- Industrial noise.

- Transport noise.

- Neighborhood noise.

1. Industrial Noise

Highly intense sound or noise pollution is caused by many machines. There exists a long list of sources of noise pollution including different machines of numerous factories, industries and mills.

Industrial noise, particularly from mechanical saws and pneumatic dwell is unbearable and nuisance to public.

Example: In the start industry, the workers near the heavy industrial blowers are exposed to 112 dB for eight hours and suffer from the occupational pollution.

2. Transport Noise

The main noise comes from transport. It mainly includes road traffic noise, rail traffic noise and air craft noise. The number of road vehicles like motors, scooters, cars and particularly the diesel engine vehicles have increased enormously in recent years.

A survey conducted in metropolitan cities has shown that noise level on Delhi, Bombay and Calcutta is as high as 90 dB. Inhabitants of cities are subjected to the most annoying form of transport noise which gradually deafen them.

3. Neighborhood Noise

This type of noise includes disturbances from the household gadgets and community. Common noise makes are musical instruments, TV, VCR, radios, transistors, telephones and loudspeakers, etc. Ever since the industrial revolution, noise in environment has been doubling every ten years.

Effects of Noise Pollution

- It causes the muscles to contract leading to nervous breakdown, tension, etc.

- Noise pollution affects human comfort, health and efficiency. It causes contraction of blood vessels, makes the skin pale, and leads to excessive secretion of the adrenalin hormone into blood stream which is responsible for high blood pressure.

- It affects health efficiency and behaviour. It may cause damage to heart, brain, kidneys, and livers and may also produce emotional disturbances.

- The adverse reactions are coupled with a change in hormone content of blood, which in turn increase the rate of heart beat, construction of blood vessels, digestive spasms and dilation of pupil of eye.

- Recently it has been reported that blood is also thickened by excessive noises.

- In addition to the serious loss of hearing due to excessive noise, impulsive noise also causes psychological and pathological disorders.

Control Measures

1. Source Control

This may include source modification such as acoustic treatment to machine surface, design changes, limiting the operational timings and so on.

2. Transmission Path Intervention

This may include containing the source inside a sound insulating enclosure, construction of noise barrier or provision of sound absorbing materials along the path.

3. Receptor Control

This includes protection of the receivers by altering the work schedule or provision of personal protection devices such as ear plugs for operating noisy machinery. The measure may include dissipation and deflection methods.

4. Oiling

Proper oiling will reduce the noise from the machines.

Preventive Measures

Noise can be reduced by prescribing noise limits for vehicular traffic, ban on honking of horns in certain areas and creation of silent zones near the schools and hospitals and redesigning of buildings to make them noise proof. Other measures can involve reduction of traffic density in residential areas and giving preferences to the mass public transport system.

Thermal Pollution

The excess heat present in the water will pollute the normal water body when it is merged and damages the aquatic life, is called thermal pollution.

Sources of Thermal Pollution

Some of the important sources are:

- Industrial effluents.

- Wastewater from Nuclear power plants.

- Wastewater from hydroelectric power station and domestic area.

Effects of Thermal Pollution

- The dissolved oxygen content is reduced and the aquatic organisms are affected to a large extent.

- The most of the water properties are changed.

Increase in Toxicity

- Due to the raise of temperature the activity of toxic materials are increased. This will create so many problems.

- The reproductivity and other activities of fish and other aquatic organisms will be affected because of high temperature.

- Some of the activities of microorganisms are accelerated due to high temperature.

- The rate of oxygen depletion will increase in water and it will demand more oxygen.

Interference with Biological Activities

- For the activities like respiration, digestion, excretion etc., the temperature is essential for the human body.

- Changes in temperature totally disrupt the entire ecosystem.

Interference with Reproduction

- In fishes, several activities like nest building, spawning, hatching, migration and reproduction etc., depend on an optimum temperature. For instance, the temperature at which lake front will spawn is successfully at 8 -9°C.

- The warm water not only disturbs spawning but also destroys the laid eggs.

Change in Metabolic Rate

- The activities in fish like food uptake, swimming speed, respiration etc., are increased due to the rise of water temperature.

Increased Vulnerability to Disease

- Due to the rise of temperature, the bacterial disease and several pathogenic microorganisms speeded up.

Invasion of Destructive Organisms

- Invasion of ship worms into New Jersey's Oyster Creek is the best example. A few years ago ship-worms were absent from Oyster Creek because they could not survive in low temperature water.

Changes in Algae

- Some of the algae shows excessive growth due to the nutrients present in the polluted water.

- These disrupt the aquatic food chain thereby increase toxicity to fish.

Increasing the Demand of Oxygen

- Since higher temperature increases the rates of physiological processes and favors bacterial growth, the oxidation of oxygen demanding wastes will be speed up, due to high rate of oxygen depletion.

- Thus the demand of dissolved oxygen content is aggravated further.

Control of Thermal Pollution

- Control of thermal pollution is an extreme necessity, since in future its detrimental effects on aquatic ecosystem may worsen.

- To reduce the magnitude of the pollution, the outlet water can be made to lose some of its heat to the environment and then may be discharged into the water course.

The following methods can be adopted to control the high temperature caused by thermal discharges:

- Cooling ponds.

- Spray ponds.

- Cooling towers.

Cooling Ponds

After the discharge of water from the sources the hot water is stored in the cooling ponds where the excess heat is evaporated.

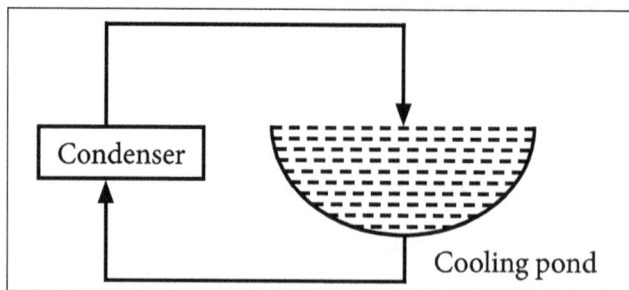

Dissipation heat by cooling ponds.

Spray Ponds

Due to spraying of water from the nozzles the excess heat is dissipated to the atmosphere.

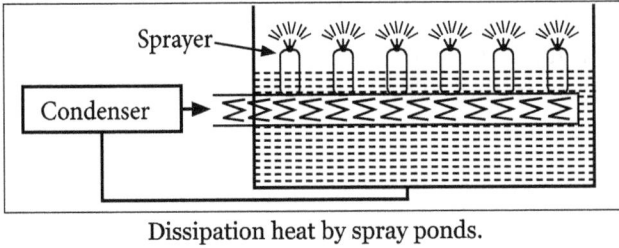

Dissipation heat by spray ponds.

Cooling Tower

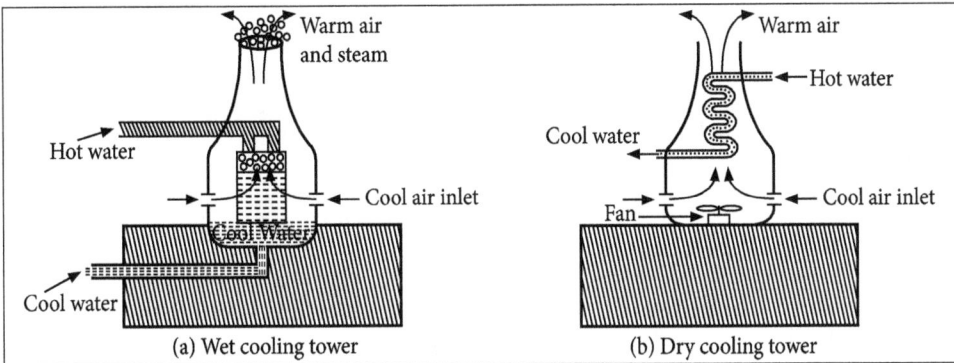

Cooling tower.

Wet Cooling Tower

In which the water is cooled in atmosphere by evaporation. The large volume of water is wasted due to evaporation.

Dry Cooling Tower

Here the cooling of water takes place in the pipeline when the water is passes with air in the pipe.

Nuclear Hazards

It is defined as the risk or danger to human health or environment posed by radiation emanating from atomic nuclei of a given substance or the possibility of an uncontrolled explosion originating from a fusion or fission reaction of atomic nuclear.

Sources of Nuclear Hazards

Sources of radioactivity are both natural and manmade.

Natural sources include:

1. Cosmic rays from outer space: The quantity depends on the altitude and latitude. It is more at higher latitudes and high altitudes.

2. Emissions from radioactive materials from Earth's crust: People have been exposed to low levels of radiation from these natural sources. But it is manmade sources which are posing a threat to mankind. The man-made sources of radioactivity are nuclear wastes produced at the time of:

- Uses of radioactive material in nuclear power plants.

- Mining and processing of radioactive ores.

- Uses of radioactive materials in nuclear weapons.

- Uses of radioactive isotopes in medical, industrial and research applications.

The greatest exposure to human beings comes from the diagnostic use of the X-rays, radioactive isotopes used as the tracers and the treatment of cancer and other ailments.

Effects of Nuclear Hazards

Effects of radioactive pollutants depend upon halflife, energy releasing capacity, rate of diffusion and rate of deposition of the contaminant. Various atmospheric conditions and climatic conditions such as temperature, wind and rainfall also determine their effect. All organisms are affected from the radiation pollution and the effects are extremely dangerous.

The effects may be somatic or genetic damage. The effects are cancer, shortening of life span and genetic effects or mutations. Some of the possible effects are listed as below:

- Exposure at low doses of radiations (100-250 rads), men do not die but begin to suffer from fatigue, nausea, vomiting and loss of hair. But recovery can be possible.

- Radiations may break the chemical bonds such as DNA in cells. This affects the genetic makeup and control mechanisms. The effects can be instantaneous, prolonged or delayed types. Even it could be carried to future generations.

- Higher irradiation doses (10,000rads) kills the organisms by damaging the tissues of heart, brain, etc.

- Exposure at higher doses (400-500 rads), blood cells are reduced, bone marrow is affected, blood fails to clot and the irradiated person soon dies of infection and bleeding.

- Through food chain also, the radioactivity effects are experienced by man. But the most significant effect of radioactivity causes long range effects, natural resistance and fighting capacity against germs is reduced, affecting the future of man and hence the future of our civilization.

- Workers handling the radioactive wastes get slow but continuous irradiation and in course of time develop cancer of different types.

Control of Nuclear Hazards

The peaceful uses of radioactive materials are so wide and effective that modern civilization cannot go without them on the other hand, there is no cure for radiation damage. Hence, the only option against nuclear hazards is to check and prevent radioactive pollution:

- Safety measures are enforced strictly.

- Leakages from nuclear reactors, careless handling, transport and use of radioactive fuels, fission products and radioactive isotopes have to be totally stopped.

- There should be regular monitoring and quantitative analysis through frequent sampling in the risk areas.

- Waste disposal is careful, efficient and effective.

- Appropriate steps should be taken against occupational exposure.

- Preventive measures should be followed so that background radiation levels do not exceed the permissible limits.

- Safety measures should be strengthened against the nuclear accidents.

Disposal of Nuclear Wastes

Since nuclear waste can be extremely dangerous and therefore, the way in which they are to be disposed of is strictly controlled by international agreement. Since 1983, by international agreement, the disposal in the Atlantic Ocean and into the atmosphere has been banned. After processing, to recover usable material and reducing the radioactivity of the waste, disposal is made in solid form where possible. The nuclear wastes are classified into three categories such as:

1. High Level Wastes (HLW)

It have a very high radioactivity per unit volume. For example, spent nuclear fuel have to be cooled and are, therefore, stored for several decades by its producer before disposal. Since these wastes are too dangerous to be released anywhere in the biosphere, they must be contained either by converting them into inert solids and then buried deep into earth or stored in deep salt mines.

2. Medium Level Wastes (MLW)

They are solidified and are mixed with concrete in steel drums before being buried in deep mines or below the sea bed in concrete chambers.

3. Low Liquid Wastes (LLW)

They are disposed of in steel drums in concrete lined trenches in designated sites.

4.2   Solid Waste Management: Causes and Effects of Urban and Industrial Wastes

Sanitary landfill the collecting, treating and disposing of the solid material that is discarded because it has served its purpose or it is no longer useful. Improper disposal of municipal solid waste can create unsanitary conditions and these conditions in turn can lead to the pollution of the environment and to outbreaks of vector-borne disease that is, diseases spread by rodents and insects.

The tasks of solid-waste management present complex technical challenges. They also pose a wide variety of administrative, economic and social problems which must be managed and solved.

Solid waste are broadly be classified into two categories. According to the Indian MSW, Rules 2000 "Municipal Solid Waste" includes both commercial and domestic wastes generated in a municipal or notified area in either solid or semisolid form excluding the industrial hazardous wastes but including treated biomedical wastes. Solid waste also includes the hazardous waste generated by various industries.

Municipal Solid Waste (MSW) can be further classified into biodegradable waste recyclable materials (like paper, glass, bottles, metals and certain plastics) and domestic hazardous waste (like medication, chemicals, light bulbs and batteries).

Solid waste management is greatest challenge that is faced by many countries around the globe. Inadequate collection, recycling or treatment and uncontrolled disposal of the waste in dumps can lead to severe hazards, like health risks and environmental pollution.

The management of the solid waste typically involves its collection, transport, processing and recycling or disposal.

Collection includes the gathering of solid waste and the recyclable materials and the transport of these materials, after collection to a location where the collection vehicle is emptied. This location can be a material processing facility, a transfer station or a landfill disposal site.

Waste disposal now-a-days is done primarily by the land filling or closure of existing dump sites. Modern sanitary landfills are not dumps they are engineered facilities used for disposing of solid wastes on land without creating hazards to public health or safety, such as the breeding of insects and the contamination of ground water.

The examples of solid waste management are given below:

- Nuclear, thermal, plastic, medical - industrial solid waste.

- Agriculture, domestic - urban solid waste.

- e-waste.

Rapid population growth and urbanization in developing countries has led to people generating enormous quantities of solid waste and consequent environmental degradation. The waste is normally disposed in open dumps creating nuisance and an environmental degradation.

Solid wastes cause a major risk to public health and the environment. Management of solid wastes is important in order to minimize the adverse effects posed by their indiscriminate disposal.

Types of Solid Wastes

Depending on the nature of origin, solid wastes are classified into:

- Urban or Municipal wastes.

- Industrial wastes.

- Hazardous wastes.

Sources of Urban Wastes

Urban wastes include the following wastes:

- Domestic wastes: It containing a variety of materials thrown out from homes.

  Example: Food waste, Cloth, Waste paper, Glass bottles, Polythene bags, Waste metals, etc.

- Commercial wastes: It includes the wastes that comes out from shops, markets, hotels, offices, institutions, etc.

  Example: Waste paper, packaging material, cans, bottle, polythene bags, etc.

- Construction wastes: It includes wastes of construction materials.

  Example: Wood, Concrete, Debris, etc.

- Biomedical wastes: It includes mostly waste organic materials.

  Example: Anatomical wastes, Infectious wastes, etc.

Classification of Urban Wastes

Biodegradable Wastes

Those wastes that can be degraded by the micro-organisms are called biodegradable wastes.

Example: Food, vegetables, tea leaves, dry leaves, etc.

Non-biodegradable Wastes

An urban solid waste materials that cannot be degraded by microorganisms are called no biodegradable wastes.

Example: Polythene bags, scrap materials, glass bottles, etc.

Sources of Industrial Wastes

The main source of industrial wastes are chemical industries, metal and mineral processing industries.

Example: Thermal power plants produces fly ash in large quantities, Nuclear plants generated radioactive wastes, Chemical Industries produces large quantities of hazardous and toxic materials and Other industries produce packing materials, scrap metals organic wastes, rubbish, acid, alkali, rubber, plastic, dyes, paper, glass, wood, oils, paints, etc.

Causes of Industrial Pollution

- Lack of Policies to Control Pollution: Lack of effective policies and poor enforcement drive allowed many industries to bypass laws made by pollution control board which resulted in mass scale pollution that affected lives of many people.

- Unplanned Industrial Growth: In most industrial townships, unplanned growth took place wherein those companies flouted rules and norms and polluted the environment with both air and water pollution.

- Use of Outdated Technologies: Most industries still rely on old technologies to produce products that generate large amount of waste. To avoid high cost and expenditure, many companies still make use of traditional technologies to produce high end products.

- Presence of Large Number of Small Scale Industries: Many small scale industries and factories that don't have enough capital and rely on government grants to run their day-to-day businesses often escape environment regulations and release large amount of toxic gases in the atmosphere.

- Inefficient Waste Disposal: Water pollution and soil pollution are often caused directly due to inefficiency in disposal of waste. Long term exposure to polluted air and water causes chronic health problems, making the issue of industrial pollution into a severe one. It also lowers the air quality in surrounding areas which causes many respiratory disorders.

- Leaching of Resources From Our Natural World: Industries do require large amount of raw material to make them into finished products. This requires extraction of minerals from beneath the earth. The extracted minerals can cause soil pollution when spilled on the earth. Leaks from vessels can cause oil spills that may prove harmful for marine life.

Effects of Industrial Pollution

1. Water Pollution

The effects of industrial pollution are far reaching and liable to affect the ecosystem for many years to come. Most industries require large amounts of water for their work. When involved in a series of processes, the water comes into contact with heavy metals, harmful chemicals, radioactive waste and even organic sludge.

These are either dumped into open oceans or rivers. As a result, many of our water sources have high amount of industrial waste in them which seriously impacts the health of our ecosystem. The same water is then used by farmers for irrigation purpose which affects the quality of food that is produced.

Water pollution has already rendered many ground water resources useless for humans and wildlife. It can at best be recycled for further usage in industries.

2. Soil Pollution

Soil pollution is creating problems in agriculture and destroying local vegetation. It also causes chronic health issues to the people that come in contact with such soil on a daily basis.

3. Air Pollution

Air pollution has led to a steep increase in various illnesses and it continues to affect us on a daily basis. With so many small, mid and large scale industries coming up, air pollution has taken toll on the health of the people and the environment.

4. Wildlife Extinction

By and large, the issue of industrial pollution shows us that it causes natural rhythms and patterns to fail, meaning that the wildlife is getting affected in a severe manner.

Habitats are being lost, species are becoming extinct and it is harder for the environment to recover from each natural disaster. Major industrial accidents like oil spills, fires, leak of radioactive material and damage to property are harder to clean-up as they have a higher impact in a shorter span of time.

5. Global Warming

With the rise in industrial pollution, global warming has been increasing at a steady pace. Smoke and greenhouse gases are being released by industries into the air which causes increase in global warming. Melting of glaciers, extinction of polar beers, floods, tsunamis, hurricanes are few of the effects of global warming.

Effect of Improper Solid Waste Management

Due to improper disposal of municipal solid waste on the roads and immediate surroundings, biodegradable materials undergo decomposition producing foul smell and become a breeding ground for disease vectors. Industrial solid wastes are source for the toxic metals and hazardous wastes that affect the soil characteristics and productivity of the soils when they are dumped on the soil.

Toxic substances may percolate into the ground and contaminate the groundwater. Burning of industrial or domestic wastes produce furans, dioxins and polychlorinated biphenyls which are harmful to human beings. Solid waste management involves waste generation, mode of collection, transportation, segregation of wastes and disposal techniques.

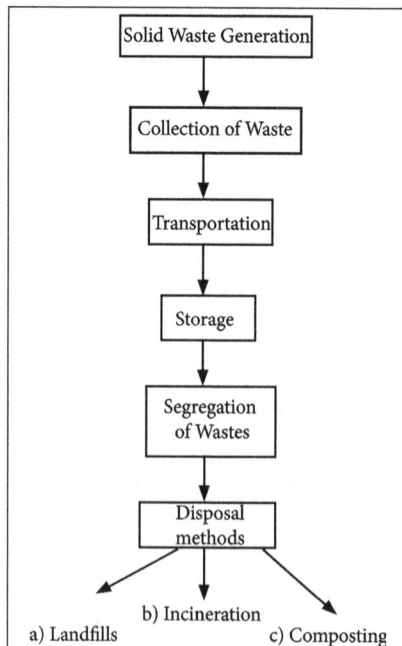

Control measures.

4.2.1 Control Measures of Urban and Industrial Wastes

An integrated waste management strategy includes three main components such as:

- Source reduction.

- Recycling.

- Disposal.

Source Reduction

Source reduction is a fundamental way to reduce waste. This can be done by minimizing the material when making a product, reuse of the products on site, designing products or packaging which helps to reduce their quantity. On an individual level we can reduce the use of unnecessary items while shopping, buy items which requires and has minimal packaging, avoid buying disposable items and also avoid using plastic carry bags.

Recycling

Recycling is reusing components of the waste that may have some economic value. Recycling has many visible benefits such as conservation of resources reduction in the energy used during manufacture and reducing the pollution levels. Some materials such as aluminum and steel can be recycled many times. Metal, paper, glass and plastics are recyclable.

Mining of the new aluminum is very expensive and hence recycled aluminum has a strong market and plays a significant role in the aluminum industry. Paper recycling can also help in preserving the forests as it needs about 17 trees to make one ton of paper. Crushed glass also known as cullet, reduces the energy that is required to manufacture new glass by 50 percent. Cullet lowers the temperature requirement of glass-making process thus conserving energy and reducing air pollution.

However even if recycling is a viable alternative, it presents several problems. Problems associated with recycling are either technical or economical. Plastics are difficult to recycle because of different types of polymer resins used in their production. Since each type has its own chemical makeup different plastics cannot be recycled together. Thus separation of different plastics before recycling is necessary.

Similarly in recycled paper the fibers are weakened and is difficult to control the colour of recycled product. The recycled paper is banned for use in food containers to prevent the possibility of contamination. It very often costs less to transport raw paper pulp than scrap paper. Collection, sorting and transport account for about 90 percent of the cost of paper recycling.

The processes of pulping, drinking and screening of wastepaper are generally more expensive than making paper from virgin wood or cellulose fibers. Very often thus recycled paper is more expensive than virgin paper. Although as technology improves the cost will come down.

Disposal

Disposal of solid waste can be done most commonly through a sanitary landfill or through incineration. A modern sanitary landfill is a depression in an impermeable soil layer that is lined with an impermeable membrane. The three key characteristics of a municipal sanitary landfill that distinguish it from an open dump are:

- Solid waste is placed in a suitably selected and prepared landfill site in a carefully prescribed manner.

- The waste material is spread out and compacted with appropriate heavy machinery.

Discarding Wastes

The following methods are adopted for discarding solid wastes:

- Landfill.

- Incineration.

- Composting.

Landfill

Solid wastes are placed in a sanitary landfill in which alternate layers of 80 cm covered with selected earth fill of 20 cm thickness. After 2-3 years solid waste volume shrinks by 25-30% and land is used for parks, roads and small buildings. This is the most common and cheapest method of waste disposal and is mostly employed in Indian cities.

Advantages

- It is simple and economical.

- Segregation of wastes is not required.

- Landfilled areas can be reclaimed and used for other purposes.

- Converts low lying, marshy waste-land into useful areas.

- Natural resources are returned to soil and recycled.

Disadvantages

- Large area is required.

- Land availability is away from the town, transportation costs are high.

- Leads to bad odour if landfill is not properly managed.

- Land filled areas will be sources of mosquitoes and flies requiring application of insecticides and pesticides at regular intervals.

- Causes fire hazard due to formation of methane in wet weather.

Incineration

It is a hygienic way of disposing solid waste. It is suitable if waste contains more hazardous material and organic content. It is a thermal process and very effective for detoxification of all combustible pathogens. It is expensive when compared to composting or land filling.

In this method, municipal solid wastes are burnt in a furnace called incinerator. Combustible substances such as rubbish organisms and garbage, dead combustible matter such as glass, porcelain and metals are separated before feeding to incinerators.

The non-combustible materials can be left out for recycling and reuse. The leftover ashes and clinkers may account for about 10 to 20% which need further disposal by sanitary landfill or some other means.

The heat produced in the incinerator during burning of refuse is used in the form of the steam power for generation of electricity through turbines. The municipal solid waste is generally wet and has a high calorific value. Therefore, it has to be dried first before burning.

Waste is dried in a preheater from where it is then taken to a large incinerating furnace known as the "destructor" which can incinerate about 100 to 150 tonnes per hour. Temperature normally maintained in the combustion chamber is about 700°C which may be increased to 1000°C when electricity is to be generated.

Advantages

- Residue is only 20-25% original and can be used as clinker after treatment.

- Requires very little space.

- Cost of transportation is not high, then if the incinerator is located within the city limits.

- Safest from hygienic point of view.

- An incinerator plant of 3000 tonnes per day capacity can generate 3MW of power.

Disadvantages

- Its capital and operating cost is high.

- Operation needs skilled personnel.

- Formation of smoke, dust and ashes needs further disposal and that may cause air pollution.

Composting

It is another popular method practiced in many cities in our country. In this method, bulk organic waste is converted into fertilizer by biological action.

Advantages

- Manure added to soil increases water retention and ion-exchange capacity of soil.

- This method can be used to treat several industrial solid wastes.

- Manure can be sold thereby reducing cost of disposing waste.

- Recycling can be done.

Disadvantages

- Non-consumables have to be disposed separately.

- The technology has not caught-up with the farmers and hence does not have an assured market.

4.3   Role of an Individual in Prevention of Pollution

Environmental pollution cannot be prevented and removed. The proper implementation and particularly the individual participation are the important aspects which should be given due importance.

The individual participation is useful in law making processes and restraining the pollution activities and thereby the public participation plays a major role in the effective environmental management. A small effect is made by each individual at his own place will have pronounced effect at the global level. It is suitably said "Think globally act locally."

Each individual should change his or her life style in a such a way as to reduce environmental pollution.

Individual Participation

- Help more in pollution prevention than pollution control.
- Plant more trees.
- Purchase recyclable, recycled and environmentally safe products.
- Use water, energy and other resources efficiently.
- Use natural gas than coal.
- Use CFC free refrigerators.
- Increase use of renewable resources.
- Reduce deforestation.
- Use office machines in well ventilated areas.
- Remove No_x from motor vehicular exhaust.

4.3.1 Pollution Case Studies

The Ganga, India

There is a universal reverence to water in almost all of the major religions of the world. Most religious beliefs involve some ceremonial use of "holy" water. The belief in its known historical and unknown mythological origins, the purity of such water and the inaccessibility of remote sources, elevate its importance even further.

In India, the river Ganga occupies a unique position in the cultural ethos of India. Legend says that the river has descended from the Heaven on earth as a result of the long and arduous prayers of King Bhagirathi for the salvation of his deceased ancestors. From times immemorial, the Ganga has been India's river of faith, devotion and worship. Millions of Hindus accept its water as sacred. Even today, people carry the treasured Ganga water all over India and abroad because it is a "holy" water and known for its "curative" properties.

Ganga River

The Ganga rises on the southern slopes of the Himalayan ranges from the Gangotri glacier at 4,000m above mean sea level. It flows swiftly for 250km in the mountains, descending steeply to an elevation of 288m above mean sea level. In the Himalayan region, Bhagirathi is joined by the tributaries Mandakini and Alaknanda to form the river Ganga.

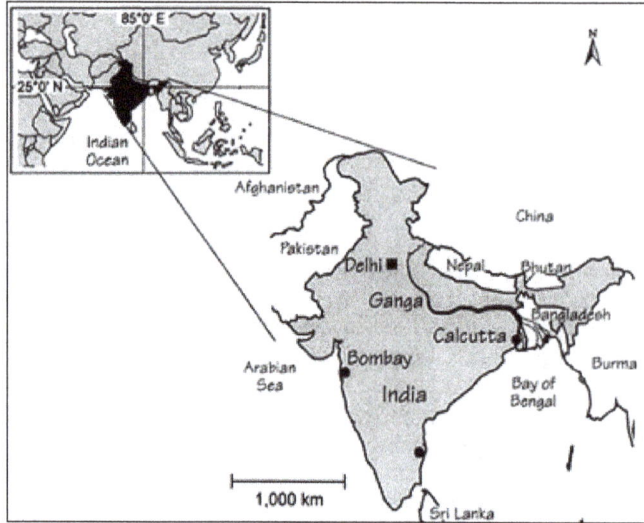

Location map of India showing the Ganga River.

After entering the plains at Haridwar, it winds its way to the Bay of Bengal, covering 2,500km through the provinces of Bihar, Uttar Pradesh and West Bengal. In the plains, it is joined by Ramganga, Yamuna, Sai, Gomti, Ghaghara, Sone, Gandak, Kosi and Damodar along with many other smaller rivers.

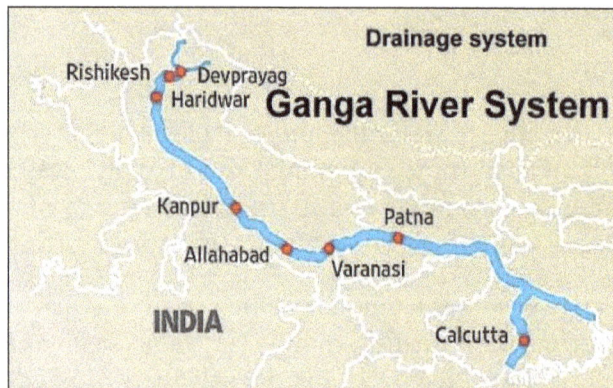

Map of India showing the route of the Ganga River.

The purity of the water depends on the velocity and the dilution capacity of the river. A large part of the flow of the Ganga is abstracted for irrigation just as it enters the plains at Hardiwar. The Ganga receives over 60 per cent of its discharge from its tributaries. The contribution of most of the tributaries to the pollution load is small, except from the Damador, Gomti and Yamuna rivers, for which separate action programs were already started under Phase II of "The National Rivers Conservation Plan".

Exploitation

In the recent past, due to rapid progress in communications and commerce, there has been a swift increase in the urban areas along the river Ganga, As a result the river is no longer only a source of water but is also a channel, receiving and transporting urban wastes away from the towns. Today, 1/3rd of the country's urban population lives in the towns of the Ganga basin. Out of the 2,300 towns in the country, 692 are located in this basin and of these, 100 are located along the river bank itself.

The belief the Ganga river is "holy" has not, however, prevented over-use, abuse and pollution of the river. Due to over-abstraction of water for irrigation in the upper regions of the river, the dry weather flow has been reduced to a trickle. Rampant deforestation in the last few decades results in topsoil erosion in the catchment area, has increased silt deposits which, in turn, raise the river bed and lead to devastating floods in the rainy season and stagnant flow in the dry season.

Along the main river course, there are 25 towns with a population of more than 100,000 and about another 23 towns with populations above 50,000. In addition there are 50 smaller towns with populations above 20,000. The natural assimilative capacity of the river is severely stressed.

Sources of Pollution of the Ganga

The principal sources of pollution of the Ganga river can be characterized as follows:

- Solid garbage thrown directly into the river.

- Domestic and industrial wastes: It has been estimated that about $1.4 \times 10^6 m^3 d^{-1}$ of the domestic wastewater and $0.26 \times 10^6 m^3 d^{-1}$ of the industrial sewage are going into the river.

- Animal carcasses and half-burned and unburned human corpses are thrown into the river.

- Non-point sources of pollution from agricultural run-off containing residues of harmful pesticides and fertilizers.

- Mass bathing and ritualistic practices.

- Defecation on the banks by low-income people.

Ganga Action Plan

Scientific Awareness

There are 14 major river basins in India with natural waters which are being used for human and developmental activities. These activities contributes to the pollution loads of these river basins. Of these river basins, the Ganga sustains the largest population.

The Central Pollution Control Board (CPCB), which is India's national body for monitoring the environmental pollution, undertook a comprehensive scientific survey in 1981-82 to classify the river waters according to their designated best uses. This report was the first systematic document that formed the basis of the Ganga Action Plan (GAP).

It detailed the land-use patterns, domestic and industrial pollution loads, fertilizer and pesticide use, hydrological aspects and river classifications. This inventory of pollution was used by the Department of Environment in the year 1984 when formulating a policy document.

Realizing the need for urgent intervention of the Central Ganga Authority was set up in 1985 under the chairmanship of the Prime Minister. The Ganga Project Directorate was established in June 1985 as a national body operating within the National Ministry of Environment and Forest.

The GPD serve as the secretariat to the CGA and also as the Apex Nodal Agency for implementation. It was set up to co-ordinate the different ministries involved and to administer funds for this 100 per cent centrally-sponsored plan.

The programme was perceived as a once-off investment providing demonstrable effects on the river water quality. The execution of works and its subsequent operation and management (O&M) were the responsibility of the state governments, under the supervision of GPD.

GPD was to remain in place until the GAP was completed. The plan was formally launched on 14 June 1986. The main thrust was to divert and intercept the wastes from urban settlements away from the river.

Treatment and economical uses of waste were made an integral part of the plan. It was realized that the comprehensive co-ordinated research would have to be conducted on the following aspects of Ganga:

- A more rational plan for the use of the resources of the Ganga for animal husbandry, agriculture, forests, fisheries, etc.

- The sources and nature of the pollution.

- The possible revival of the inland water transport facilities of the Ganga together with the tributaries and distributaries.

- The demographic, cultural and human settlements on the banks of the river.

One outcome of this initiative was a multidisciplinary study of the river in which 14 universities located in the basin participated in a well-co-ordinated, integrated research programme. This was one of the largest endeavors, involving several hundred scientists, ever undertaken in the country and was funded under GAP. The resultant report is a unique, integrated profile of the river.

GAP was the first step in river water quality management. Its mandate was limited to quick and effective, but sustainable, interventions to contain the damage. The studies carried out by the CPCB in 1981-82 revealed that pollution of the Ganga was increasing but had not assumed serious proportions, except at the certain main towns on the river such as industrial Kanpur and Calcutta on the Hoogly, together with a few other towns.

The causative factors responsible for these situations were targeted for swift and effective control measures. This strategy was adopted for urgent implementation during the first phase of the plan under which only 25 towns identified on the main river were to be included.

The studies had revealed that:

- 8 percent of the municipal sewage was from the 25 Class I towns on the main river.

- 75 percent of the pollution load was from untreated municipal sewage.

- All the industries accounted for only 25 percent of the total pollution.

- Only a few of these cities had sewage treatment facilities.

Attainable Objectives

The broad aim of the GAP was to reduce the pollution, clean the river and to restore the water quality at least to Class B (i.e. bathing quality: 3 mg l^{-1} BOD and 5 mg l^{-1} dissolved oxygen). This was considered as a feasible objective and because a unique and distinguishing feature of Ganga was its widespread use for ritualistic mass bathing.

The multipronged objectives were to improve the water quality by controlling the municipal and industrial wastes. The long-term objectives were to improve the environmental conditions along the river by suitably reducing all the polluting influences at source.

Prior to the creation of the GAP, responsibilities for pollution of the river were not clearly demarcated between the various government agencies. The pollutants reaching Ganga from most point sources did not mix well in the river, due to the sluggish water currents and as a result such pollution often lingered along the embankments where people bathed and took water for domestic use.

Strategy

The GAP had a multipronged strategy to improve the river water quality. It was fully financed by central Government, with the assets created by the central Government to be used and maintained by the state governments. The main thrust of the plan was targeted to control all municipal and industrial wastes.

All the possible point and non-point sources of pollution were identified. The control of urban non-point sources was also tackled by direct interventions from the project funds. The control of point sources of urban municipal wastes for the 25 Class I towns on the main river was initiated from 100percent centrally-invested project funds.

The control of non-point source agricultural run-off was undertaken in a phased manner by the Ministry of Agriculture, principally by reducing the use of fertilizer and pesticides. The control of point sources of industrial wastes was done by applying the polluter-pays-principle.

A total of 261 sub-projects were sought for implementation in 25 Class I river front towns. This would eventually involve a financial outlay of Rs.4,680 million (Indian Rupees), equivalent to about US $156 million.

More than 95 percent of the programme has been completed and remaining sub-projects are in various stages of completion. The resultant improvement in the river water quality is hotly debated in the media by certain non-governmental organizations (NGOs).

The success of the programme can be gauged by fact that Phase II of the plan, covering some of the tributaries, has already been launched by the Government. In addition, the earlier action plan has now evolved further to cover all the other major national river-basins in India, including a few lakes and is known as the "National Rivers Conservation Plan".

Domestic Waste

The major problem of pollution from the domestic municipal sewage (1.34×10^6 m^{3}d^{-1}) arising from the 25 selected towns were handled directly by financing the creation of facilities for diversion, interception and treatment of wastewater and also by preventing the other city wastes from entering the river.

Out of the 1.34×10^6 m^{3}d^{-1} of sewage assessed to be generated, 0.873×10^6 m^{3}d^{-1} was intercepted by laying 370 km of trunk sewers with 129 pumping stations as part of 88 sub-projects.

The activities of various sub-projects can be summarized as follows:

Approach to river water quality improvement	Number of schemes
Sewage treatment plants	35

Interception and diversion of municipal wastewater	88
Electric crematoriums	28
Low-cost sanitation complexes	43
River front facilities for bathing	35
Others (e.g. biological conservation of aquatic species, river quality monitoring)	32
Total	261

A total of 248 of these schemes have already been commissioned and the remaining are due to be completed by 1998.

Industrial Waste

About 100 industries were identified on the main river itself. Sixty-eight of these were considered grossly polluting and were discharging 260×10^3 m³d⁻¹ of wastewater into the river. Under the Water Act 1974 and Environment Act 1986, 55 industrial units (generating 232×10^3m³d⁻¹) out of the total of 68 grossly polluting industrial units complied and installed effluent treatment plants.

In addition, two others have treatment plants under construction and currently one unit does not have a treatment plant. Legal proceedings have been taken against the remaining 12 industrial units which were closed down for non-compliance.

Applied Research

The Action Plan stressed the importance of applied research projects and many universities and reputable organizations were supported with grants for projects carrying out studies and observations which would have a direct bearing on the Action Plan.

Some of the prominent subjects were sewage-fed pisciculture, PC-based software modelling, bio conservation in Bihar, conservation of fish in upper river reaches, using treated sewage for irrigation, monitoring of pesticides and rehabilitation of turtles.

Some of the ongoing research projects include land application of untreated sewage for tree plantations, disinfection of treated sewage by UV radiation, aquaculture for sewage treatment and disinfection of treated sewage by Gamma radiation. All the presently available research results are being consolidated for easy access by creation of a data base by the Indian National Scientific Documentation Centre.

Integrated Improvements of Urban Environments

The GAP also covered very wide and diverse activities, such as conservation of aquatic species (gangetic dolphin), protection of natural habitats and creating riverine sanctuaries.

It also included components for building stepped terraces on the sloped river banks for ritualistic mass-bathing (128 locations), landscaping river frontage (35 schemes), development of public facilities, improving sanitation along the river frontage (2,760 complexes), improved approach roads and lighting on the river frontage.

Public Participation

The pollution of the river, although classified as environmental, was the direct outcome of the deeper social problem emerging from long-term public diffidence and apathy, indifference and lack of public awareness, education and social values and above all from poverty.

In recognition of necessity of the involvement of the people for sustainability and success of the Action Plan, importance was given for generating awareness through intensive publicity campaigns using the audio visual approaches, press and electronic media, leaflets and hoardings as well as organizing public programs for spreading the message effectively.

In spite of full financial support from the project and heavy involvement of about 39 well known NGOs to organize these activities, the programme had only limited public impact and even received some criticism. Other similar awareness-generating programs involving school children from many schools in the project towns were received with greater enthusiasm.

Operation and Maintenance

The enduring success of the pollution abatement works under the GAP is essential for sustainability. Most of these works were carried out by the same agencies which were eventually responsible for maintaining them as part of their primary functions, such as the municipality, the city development authority or the irrigation and flood control department. The responsibility for subsequent O&M of these works automatically passed to these agencies.

The most crucial components for preventing the river pollution were the main pumping stations which were intercepting the sewage and diverting it to the treatment plants. These large capacity pumping stations, operating at the city level, had been built for the first time in India and it was considered that the municipalities would have adequate skilled personnel and resources to be able to manage them.

The central Government shared half of this deficit until 1997. In the broader interest of pollution control, future policies will also be similar, where the state governments undertake the responsibility for pollution control works because the local bodies are unable to bear the cost of O&M expenditures with such limited resources.

Technology Options

The choice of technology for the GAP was largely conventional based on the available options and local considerations. Consequently, the sewers and pumping stations and all the similar municipal and conservancy works were executed in each province by its own implementing agencies, according to their customary practices but within the commonly prescribed specifications.

The choice of technology for most of the large domestic wastewater treatment plants was carefully decided by a panel of experts in close consultation with those external aid agencies.

A parallel procedure was adopted in-house for all other similar projects. For all the larger sewage treatment plants, the unanimous choice was to adopt the well-accepted activated sludge process. Some of these new and simpler technologies, with their low-cost advantages, will emerge as the large-scale future solution to India's sanitation problems.

Implementation Problems

The implementation of a project of this magnitude over the entire 2,500 km stretch of the river, covering 25 towns and crossing three different provinces, which could be achieved by delegating the actual implementation to the state government agencies which had the appropriate capabilities. The state governments also undertook the responsibility of subsequently operating the assets being created under the programme and maintaining them.

The overall inter-agency co-ordination was done by the GPD through the state governments. The defined project objectives were ensured by the GPD through appraisal of each project component submitted by the implementing agency.

However, the involvement of aid agencies, with their associated mandatory procedures, also added to the complexities of decision-making, especially in the large STP projects. Of the original 261 sub-projects, 95 per cent are now complete and functioning satisfactorily.

River Water Quality Monitoring

Right from its inception in 1986, GAP started a very comprehensive water quality monitoring programme by obtaining the data from 27 monitoring stations. Most of these river water quality monitoring stations were already existed under other programs and only required strengthening.

Technical help was also received for a small part of this programme from the Overseas Development Agency of UK in the form of some automatic water quality monitoring stations, the associated modeling software, training and some hardware.

Monitoring programme is being run on a permanent basis using the infrastructure of other agencies such as CPCB and Central Water Commission (CWC) to monitor the data from 16 stations. To evaluate the result of this programme, an independent study of the water quality has also been awarded to separate the universities for different regional stretches of the river.

Future

Apart from the visible improvement in the water quality, awareness generated by the project is an indicator of its success. It has resulted in the expansion of the programme over the entire Ganga basin to cover the other polluted tributaries. The GAP has further evolved to cover all the polluted stretches of the major national rivers including a few lakes. Considering the huge costs involved the central and state governments have agreed in principle to each share half of the costs of the projects under the "National Rivers Action Plan".

Conclusions

The GAP is a successful example of timely action due to environmental awareness at the governmental level. Even more than this, it exhibits the achievement potential which is attainable by "political will". It is a model which is constantly being upgraded and improved in other river pollution prevention projects. This may be due to higher water consumption, less nutritious dietary habits, higher grit loads, fewer sewer connections, insufficient flows and stagnation leading to bio-degradation of volatile fractions in the pipes themselves.

The most important lesson learned was the need for control of pathogenic contamination in treated effluent. This could not be tackled before because of the lack of safe and suitable technology but is now being attempted through research and by developing a suitable indigenous technology, which should not impart traces of any harmful residues in the treated effluent detrimental to the aquatic life.

4.4   Disaster Management

The term disaster is derived from a French word meaning evil or bad star. The progress of science is amazing today but still mankind is unable to come to grips with natural disasters.

Minimizing the adverse effects of natural and man-made disasters by adopting suitable means (through warnings in particular) is called as disaster management. It is better to take necessary actions before a disaster rather than attempting to save lives and property after the disaster has already occurred.

Types of Disasters classified on their origin are:

- Natural-disaster.

- Manmade-disaster.

Natural disaster is further classified as wind related e.g. storm, cyclone, tornado, hurricane etc. Water related e.g. floods flash floods, excessive rain etc. Geological e.g. earthquakes, landslides, volcanic eruptions etc. Climatic disasters e.g. drought, famines, under sea earthquake resulting in very high sea waves invading the coasts at very high speed called Tsunamis, etc.

Man-made disasters are nuclear explosions, industrial accidents, fires of various kinds, accidents of trains/aircraft etc.

Floods

Major floods hit India, Nepal, Bangladesh and much of East Asia, every year killing thousands of people and animals. The Catastrophic Yangtze river flood in 1998, claimed more than 3000 lives. The Central Water Commission (CWC) in India has flood forecasting systems with 157 flood forecasting centers covering 62 interstate river basins.

These centers in collaboration with Indian Metrological Department (IMD) monitor rainfall situations and water level in reservoirs. With this information, the CWC issues flood forecast and warning about floods.

Earthquakes

A UNESCO study shows that 10,000 people die each year from earthquakes. Earthquakes in Turkey in 1999 have claimed 45,000 lives. An earthquake in Taiwan claimed 1,600 lives. In India, Uttaranchal is very sensitive zone from the point of view of earthquakes. An earthquake at Chamoli in 1999 caused 200 deaths.

Similarly an earthquake in Gujarat on 26th January, 2001 claimed thousands of lives, following measures should be taken for protection against earthquakes:

- People should be well aware and educated against the earth-quake danger.

- Canvassing should be done again and again for constructing earthquake resistant houses in a specific manner in earthquake prone areas for safety.

- Construction of multistoried building should be stopped at a hilly area or it should be not more than 12 meter high. For it a strict policy and law should be framed and be followed.

Cyclone

In 1991, Bangladesh suffered 233 km/hr cyclone winds as well as major inland flooding in which 1,40, 000 people died. Earlier in 1970 a cyclone in Bangladesh claimed

4,00,000 lives. In India cyclone forms a regular feature in Orissa. Following efforts has been taken for protection against cyclones:

- Cyclone shelters should be constructed in cyclone prone areas. Around 1,200 cyclone shelters have been constructed in the coastal regions.

- Efforts have been made to forest the coastal areas to break winds.

- Attempts are now being made to link development programmes with disaster mitigation efforts.

- Some cyclone resistant houses have been made.

- A cropping strategy has been evolved, keeping in view cyclone seasons to reduce loss of crops.

- Technical assistance regarding monitoring of cyclone detection comes from India Metrological department which works through 10 cyclone detection radars located on coasts. A geo-stationary satellite (INSAT-113) monitors cyclone movements.

Landslides

Shifting of land pieces due to natural calamities such as earthquake and excessive rainfall are called landslides:

- Causes and effects: The main causes of landslides are earthquakes and deforestation. Other causes includes blasting operations for making tunnels, dams and roads etc. Landslides results in large scale destruction of life and property.

- Management: Extensive forestation should be done to prevent the occurrence of landslides. Better means of communication should be adopted to avoid panic among people. Concerted efforts should be made for people and supply of essential items like food, medicine, clothing etc. for affected people.

Chapter 5

Social Issues and Environment

5.1 Sustainable Development

Being a rational being man is researching, collecting information and moving upward in the aura of development from the very onset of the civilization. The hindrances and hassles that he feels in his line of Action towards attaining the materialistic goals, be quashed with no qualm.

In this process many a vistas of nature have so far been suffered by the materialism craze. As a remedy to this nasty cult, the concept of sustainable development has been forwarded. Oxford advanced learner's Dictionary defines the word sustainable (adj) as,

- Involving the use of natural products and energy in a way that does not harm the environment: sustainable forest management, an environmentally sustainable society,

- That can continue or be continued for a long time.

Thus we can say that sustainable development is a regularly maintained development. In consuming the natural resources sustainable development provides for using the resources in such a way that it does not get exhausted completely but lingers to get renewed for the coming generation.

An awareness towards sustainable development on world map can be seen in the last quarter of 20th century. It is in 1972, when United Nations conference on Human Environment discussed the issue at Stockholm, a serious deliberation was made among the delegates of 113 countries.

They decided to frame policies and legislation towards preserving the environment and bringing about a sustainable development. The Stockholm declaration is known as the Magna Carts of Environment. After twenty years the U.N. Conference on Environment and Development was convened at Rio-de-Janeiro, Brazil (in June 1992). Rio-de-Janeiro Conference also known as Earth Summit formulated Agenda 21, a blue print for making development socially, economically and environmentally sustainable.

Marching towards the goal of sustainable development world summit on Sustainable Development was convened during August/ September 2002 at Johannesburg. It is the biggest summit convened so far to discuss the future of the earth and its inhabitants. This summit analysed the Agenda 21 and agreed on plan of implementation and declaration of political commitment.

The United Nations World Commission on Environment and Development propagates that, sustainable development must meet the needs of the present generation without compromising on the ability of the future generation to meet their own need and aspirations.

Nature of Sustainable Development

- Sustainable development is a comprehensive way of thinking. It is beyond the limit of time and clime. It focuses the future and centres the world peace.

- Sustainable developments ban the depletion of renewable resources at faster speed than their generation.

- Sustainable development is closely linked to the carrying of an ecosystem.

- It also focuses the conservation of bio-diversity.

- Sustainable development aims at improving the quality of life.

- Sustainable development is a benevolent and altruistic concept.

- Sustainable development is a cyclic management.

Approaches to Sustainable Development

In our blind pursuit of material possessions under the guise of so called civilization we have mended the environment. Our material advancement has certainly brought creative comforts but it could not bring contentment. This life style is damaging and unsustainable to the environment. For bringing about sustainable development we can exercise the following approaches:

- Exploring the sea and water resources as a great food sources.

- Analyzing greater prospects of development in tourism and heritage sector.

- The development and promotion of non-conventional/ alternate/ new and renewable sources of energy such as solar energy, wind energy, bio-energy ocean energy etc. (Energy from ocean can be obtained in at least eight ways i.e., Ocean Thermal Energy Conversion, Wave Energy, Tidal Energy, Current Energy, Ocean Wind Energy, Salinity Gradient Energy, Ocean Geothermal Energy, Bio-Conversion Energy).

- Use of animal energy.

- Use of alternative fuels like Compressed Natural Gas (CNG), Hydrogen, Gasohol and Hydrocarbon.

- Use of bio-fertilizers and manures.

- Introducing Integrated Pest Management (IPM) and Integrated Nutrition Management (INM) in fanning system.

- Supporting Voluntary agencies in rural areas involved in time bound projects relating to application of science and technology for human Welfare.

- Strengthening, catalyzing and supporting linkages between Field Groups, Science and Technology based NGO's, Universities and Science and Technology Institution involved in R&D application of innovative solution for development of the disadvantaged and economically weaker section of society.

- To preserve and upgrade the skills of traditional artisans as natural carriers of science and technology for sustainable development.

- To develop new technologies to improve and diversify the local economy optimum utilization of the local resources and upgrade the skills and efficiency of local people.

- To spread the concept of sustainable development through education, media and other programme among the masses.

5.1.1 Urban Problems Related to Energy

- Urbanization is a process of moving population from rural areas to urban areas for improving life standards and life style through scientific and technological developments.

The energy related problems due to urbanization include:

- Pollution from coal: The use of coal pollutes the environment.

- Acid rain: Various industries are releasing harmful gases like sulphur oxides, nitrogen oxides which reacts with water or moisture in the environment produces suphuric acid.

- Pollution from vehicle: The exhausts from two-wheeler, four wheeler and other transport vehicles produces huge level of air pollution.

- Deforestation: Human needs space to live, hence this requirement is fulfilled by deforestation and building houses. Even after this, human needs wood for house furniture and timber as fuel.

- Global warming: Combustion of fossil fuels (oil, petrol, diesel, gas) produces harmful gases, which acts as greenhouse i.e. short wave and natural light can pass but traps heat radiation hence overall environment temperature rises.

- Use of electricity: Large amount of electricity is utilized for human comforts like - A/C, washing machine, refrigerator, water heater etc. Hence, electricity requirement is increasing day by day.

5.2   Water Conservation

In ancient days women used water economically as they had to cart it for their household from long distances and with much hardship. The used water was fed to the kitchen garden and not disposed off. Water was stored in small tanks called talaabs or jheels for domestic and agricultural use.

Over the years, with increase in population, the demand for more food crop and water has also increased. At the same time, growing industrialization and demand for more agricultural land has led to deforestation. But with deforestation, the surface run-off increases and the ground water table thus drops as water cannot percolate into the ground with no vegetation.

The perennial rivers are also becoming seasonal due to the lack of forest cover. With the advancement of science and technology, ground water is also constantly withdrawn to meet the water demand. Hence, the water table is continuously receding.

Efforts in this direction are made by rain water harvesting and watershed management techniques. Emphasis should also be laid on the conservation of water by employing modified techniques such as using the drip-irrigation method to water the plants near their roots in agricultural fields rather than the traditional method. Proper pipes should be used and leakages should be checked periodically. Water should be used economically and judiciously.

This can be easily achieved, by using water from a bucket rather than using it directly from the tap and also utilizing used water (water used after washing vegetables, cereals and clothes) for the kitchen garden, cleaning floors and also by collecting rain water in buckets during the rainy season.

The surface water bodies are also responsible for recharging the ground water table of the area. Thus, the need of the hour is to recharge the aquifers, thereby recharging surface water bodies and hence, increasing the water table. We can also enrich the environment by maintaining some ground covered with vegetation, around our house. This will allow for easy percolation of rain water.

The sustainable use of rain water can only lower the demand for ground water. It can also augment local water supply through the recharge route. Ground water is a dependable source of fresh water. But, its continuous consumption has resulted in the drastic lowering of the water table. The result is the drying up of surface water bodies such as tanks, wells and ponds.

5.2.1 Rain Water Harvesting

During rainy seasons lot of water gets collected and goes waste as runoff. In view of this it will be very appropriate in the present context to capture as much of the rain water as possible and at all levels and at all places. Even individual homes can capture the rain water on their roofs and store it in some underground tanks for use throughout the year. Such water can particularly be of very good quality when it comes to use for laundry purpose.

In urban areas, the construction of houses, footpaths and roads has left little exposed earth for water to soak in. In parts of the rural areas of India, floodwater quickly flows to the rivers, which then dry up soon after the rains stop. If this water can be held back, it can seep into the ground and recharge the groundwater supply. This has become a very popular method of conserving water especially in the urban areas.

Rainwater harvesting essentially means collecting rainwater on the roofs of building and storing it underground for later use. Not only does this recharging arrest ground-water depletion, it also raises the declining water table and can help augment water supply. Rainwater harvesting and artificial recharging are becoming very important issues. It is essential to stop the decline in groundwater levels, arrest sea-Water ingress, i.e., prevent sea-water from moving landward and conserve surface water run-off during the rainy season.

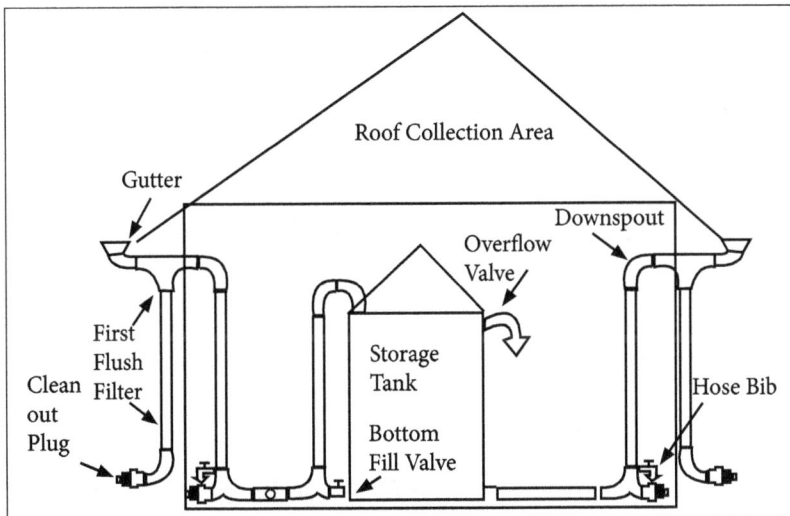

Rain water harvesting.

Town planners and civil authority in many cities in India arc introducing bylaws making rainwater harvesting compulsory in all new structures. No water or sewage connection would be given if a new building did not have provisions for rainwater harvesting. Such rules should also be implemented in all the other cities to ensure a rise in the groundwater level.

Realizing the importance of recharging groundwater, the CGWB (Central Ground Water Board) is taking steps to encourage it through rainwater harvesting in the capital and elsewhere. A number of government buildings have been asked to go in for water harvesting in Delhi and other cities of India.

Typically, rain is collected on rooftops and other surfaces and the water is carried down to where it can be used immediately or stored. We can direct water runoff from this surface to plants, trees or lawns or even to the aquifer.

Some of the benefits of rainwater harvesting are as follows:

- Increases water availability.
- Checks the declining water table.
- Is environmentally friendly.
- Improves the quality of groundwater through the dilution of fluoride, nitrate and salinity.
- Prevents soil erosion and flooding especially in urban areas.

The advantages of rain water harvesting are:

- Reduction in the use of current for pumping water.
- Migrating the effects of droughts and achieving drought proofing.
- Increasing the availability of water from well.
- Rise in ground water level.
- Minimizing the soil erosion and flood hazards.
- Upgrading the social and environmental status.
- Future generation assured of water.

5.2.2 Watershed Management

The term watershed is synonymous to the Drainage Basin, Catchment Basin or River Basin. It is the total region or area above a given point on a stream or lake that contributes water to flow at that point. It is designated by the dividing line of area from which surface streams flow in two different directions. The line separates two contiguous drainage areas.

John Wesley Powell, scientist geographer, said that a watershed is,

"Area of land, a bounded hydrologic system, within which all living things are inextricably linked by their common water course and where, as humans settled, simple logic demanded that they become part of a community."

Watersheds come in all shapes and sizes. They cross county, state and national boundaries.

Watershed.

The management of all the natural resources of a watershed to protect, maintain or improve its desired water budget, both quantity and quality, over time is called watershed management.

It involves a combination of complementary practices of land treatment and structural works to maintain or improve total yield, quality, stability of flow of surface and subsurface water and prevention of damage and loss due to excessive and uncontrolled runoff, flooding, salination and siltation.

The goal of watershed management is to plan and work toward an environmentally and economically healthy watershed that benefits all who have a stake in it.

Steps for Watershed Management

First of all, relevant information about the watershed is collected then following stages are observed to manage a watershed:

- The first stage includes uncovering concerns, gathering and analyzing information and data, defining challenges/opportunities, developing objectives and documenting data and decisions.

- The second stage includes developing a game plan for addressing the objectives, selecting the best watershed management alternative(s), listing ways (strategies) for implementing the selected alternative(s) and determining how to measure progress.

- The third stage includes implementing and evaluating efforts. The planning process requires that most interested stakeholders are involved. The map(s) and information detailing is available. Information about boundaries, terrain, water bodies, soil types, roads, land uses, recreational uses are available. There is a technical advisory team to assist.

These are some watershed management alternatives needs to be explored with the goal of selecting one or more to implement.

Example include:

Contour strips, Conservation tillage, Construction site erosion control, Filter or buffer strips, Reduced dumping of oil and/or chemicals in storm sewers, Terraces, Nutrient management, Pest management, Tree plantings, Irrigation water conservation, Home water conservation, Septic system maintenance, Alternative livestock watering sources, Roadside erosion control, Enterprise zones, Prime farmland protection, Private/rural road maintenance, Storm water management, Stream bank stabilization, Constructed wetlands, Rotational grazing, Riparian zone management.

For the purpose of watershed management, a watershed computer model may also be used. A model is a tool that watershed planners use to help them understand the cause and effect relationships within a watershed. Just like a model plane is a representation of the real plane, a watershed model can represent a real watershed.

Different types of models allow us to study different aspects. For example, one model may look at surface runoff of nutrients and pesticides while another might compare the economics of management practices.

The management team might have to use several models to address both economic and environmental concerns within our watershed. Watershed models aren't an end product, they allow us to compare differing strategies to see what might be the most economically and environmentally effective.

They provide us with information to make decisions on what alternatives to consider. The partnership of various agencies/groups must use the results of the models plus the social acceptability of those results. Only after all factors are taken into account will a decision be acceptable.

In addition to establishing a baseline prior to implementation, the management needs to consider how to evaluate the effectiveness of the plan and the progress toward the objective. This need is expensive and should be included in the Action Plan.

For instance, turbid water can be measured with a simple secchi disc, pH can be measured with a pH strip, nitrates and phosphorous can also be measured with a simple indicator strip. Wildlife can be measured by an annual count or survey. Another good barometer is the number of hunting and fishing licenses issued.

The method used for measuring change should be determined by the watershed partnership. Again, partners may want to ask for technical assistance from local conservation groups or science teachers. Regardless of the measurement, it's very important to report progress back to both the partnership and to the press.

5.3 Resettlement and Rehabilitation of People: Its Problems and Concerns

Rehabilitation of people

People are forced to evacuate from their land due to both natural and man-made disasters. Natural disasters like earthquakes, cyclones, tsunami etc.

Render thousands of people homeless and sometime even force them to move and resettle in other areas. Similarly, developmental projects like construction of dams, roads, flyovers and canals displace people from their home. Thus, resettlement refers to the process of settling again in a new area. Rehabilitation means restoration to the former state.

Objectives of Rehabilitation

The following objectives of the rehabilitation should be kept in mind before people are given an alternative site for living:

- The tribal people should be allowed to live along the lives of their own patterns and others should avoid imposing anything on them.

- They should be provided means to develop their own traditional art and culture in possible way.

- The villagers should be given the option of shifting out with the others to enable them to live community based life.

- The people displaced should get an appropriate share in the fruits of the development.

- If resettlement is not possible in the neighbor area, priority should be given to development of the irrigation facilities and supply of basic inputs for agriculture, drinking water, wells, grazing ground for the schools for the children, cattle, primary health care units and other amenities.

- Removal of poverty should be one of the objectives of rehabilitation.

- The elderly people of village should be involved in the decision making.

- The displaced people should be given employment opportunities.

- Resettlement should be in the neighborhood of their own environment.

- Villagers should be taken into confidence at every stage of implementation of the displacement and they should be educated, through public meetings, discussion about the legalities of the Land Acquisition act and other rehabilitation provisions.

Resettlement Issues

As per the World Bank estimates, nearly 10 lakhs people are displaced worldwide for a variety of reasons.

Little or No Support

Displacement mainly hits tribal and rural people who usually do not figure in the priority list of political authorities or parties.

Meager Compensation

The compensation for the land lost is often not paid, it is delayed or even if paid, is too small for both in monetary terms and social changes forced on them by these big developmental projects.

Loss of Livelihood

Displacement is not a simple incident in the lives of the displaced people. They should leave their ancestral land and forests on which they are depended for their livelihood. Many of them does not posses any skills to take up another activity or pick up any other occupation. Usually, the new land that is offered to them is of poor quality and refugees are unable to make a living.

Lack of Facilities

When people are resettled in a new area, basic infrastructure and amenities are not provided in that area. Very often, temporary camps become permanent settlements. It is also an important problem of displacement or resettlement that people have to face.

Increase in Stress

Resettlement disrupt the entire life of people. They are unable to bear the shocks of purposelessness and emptiness created in their life. The payment of compensation to the head of the family often lead to bitter quarrels over sharing of compensation amount within the family, leading to stress and even withering of family life.

Moreover, the land ownership has a certain prestige attached to it that cannot be compensated for, even after providing the new land. With the loss of property and prestige, the marriages of young people also become difficult as people from the outside villages are not willing to marry their daughters to the refugees.

Increase in Health Problems

Lack of nutrition due to the loss of agriculture and forest based livelihood, lead to the general decline in the health of the people. People are used to traditional home remedies. But the herbal remedies and plants gets submerged due to the developmental projects.

Secondary Displacement

The occupational groups which are residing outside the submergence area but depending on the area for livelihood also experience unemployment. Laborers, village artisans, petty traders etc., lose their living.

Loss of Identity

Tribal life is based on the community. The tribal are very simple people who have a lifestyle of their own. Displacement have a negative impact on their livelihood, cultural and spiritual existence in the following ways:

- Inter community marriages, folk songs, cultural functions and dances does not take place among the displaced people. When they are resettled, it is usually individual based resettlement, which ignores communal character.

- Breakup of the families and communities are the important social issues of displacement. The women suffer the most as they are deprived even a little compensation.

- Resettlement increases the poverty of tribal due to the loss of land, livelihood, food insecurity, jobs, skills etc.

- Loss of the identity of individuals and loss of connection between the people and the environment is greatest loss in the process. The indigenous knowledge that they have regarding the herbal plants and wildlife are lost.

- The laws of land acquisition does not pay attention to the idea of communal ownership of property which increases stress within the family.

- The tribal people are not familiar with market trends, prices of commodities and policies. As such, they are exploited and get alienated in modern era.

Case Studies

Displacement Due to Dams

India has been constructing dams and other hydel projects. In the last 50 years, 20 million people have been affected by the construction of such projects. The Hirakud dam displaced about 20000 people living in about 250 villages. The Bhakra Nangal dam was constructed around 1950's and displaced a number of people. Some of them could not be rehabilitated till today.

Displacement Due to Mining

Due to possibility of the accidents or sinking of the land, people have to displaced in and around the mining area. Mining take up several hectares of land thousands of people have to be evacuated. Jharia coal fields posed a problem years ago to the local residents due to the underground fire. Some 3 lakh people were to be shifted and it became a problem to find n alternative site. A huge amount of money to the tune of Rs. 115 crores has been spent to put out the fire. Still the problem persists.

Displacement in Japan Due to Nuclear Crisis

We must all be aware of current nuclear crisis in Japan where there was an explosion in three of the major reactors of Fukushima city due to tsunami. Currently, more than 2,00,000 people have been displaced from their native place and yet many are unable to find an alternative home. People were evacuated to protect them for the possible nuclear hazard and exposure. They are suffering from acute hunger as all the food supply was interrupted due to the contamination of food particles by radiation.

5.4   Environmental Ethics: Issues and Possible Solutions

Environmental ethics refers to the issues, principles and guidelines relating to human interactions with their environment. It also means that effort must be taken to protect an environment and to maintain its stability from the hazardous chemical pollutants.

The formation of an ethics of the environment has taken place within the past twenty years, even though the problems of judging human dealings with respect to the environment and in terms of morality have been known for a longer time.

The past twenty years, however, have seen the recognition of the ethics of the environment as a scientific discipline, an institutionalization. Most of all this meant an intensive development, especially in the area of its categorical apparatus and the formation of its philosophical and methodological alternatives.

In our context the mode of forming and processing problems of environmental ethics begins rather late. This has a bearing mainly on our past political development, when the ethics of the protection of the environment were subordinated to the aims and intentions of the ruling power. The protection of the environment was pushed to the edge of interests while other values, especially those of social concerns, were focused upon.

Emphasis was placed on the construction of mass settlements, the building of factories and the like, but without much consideration being given to the impact of such decisions upon the environment. The problematic of the environment consequently remained be rather in the mind of the political opposition.

In the recent times exploration in that field has noticeably improved. Proposals for environmental laws are being formulated, tens of non-governmental organizations for the protection of the environment were created, the formation of institutional assumptions for environmental ethics as a science is carried out and many articles and studies focused on the problematic of the protection of the environment are being published.

Regularly every year seminars and conferences on that theme are being organized; The time has come when it will be necessary to include this problem purposefully even into training and educational programmes. Today's opinions concerning environmental ethics are pluralistic and broadly spread.

Functions of Environment

- It is the life supporting medium for all organisms.

- It provides food, air, water and other important natural resources to the human beings.

- It disintegrates all the waste materials discharges by the modern society.

- It moderates the climatic conditions of the soil.

- A healthy economy depends on a healthy environment.

Environmental Problems

- Deforestation activities.

- Population growth and urbanization.

- Pollution due to discharge of effluent and smoke discharge from the industries.

- Water scarcity.

- Land degradation and degradation of soil fertility.

Solutions to Environmental Problems

The environmental can be protected due to the following activities:

- Reduce the waste of matter and energy resources.

- Recycle and reuse as many of our waste products and resources as possible.

- Over exploitation of natural resources must be reduced.

- Soil degradation must be minimized.

- Sustainable development is essential on, conservation of resources, harvesting of non-conventional energy and waste management.

- Biodiversity of the earth must be protected.

- Reduce population and increase the economic growth of our country.

5.5   Climate Change and Global Warming

Climate Change

The activities of the transport industry release several million tons of gases each year into the atmosphere. These include lead, carbon dioxide, methane, carbon monoxide, nitrogen oxides, nitrous oxide, chlorofluorocarbons, perfluorocarbons, silicon tetrafluoride (SF_6), benzene and volatile components, heavy metals and particulate matters.

Some of these gases, particularly nitrous oxide, also participate in depleting the stratospheric ozone layer which naturally screens the earth's surface from UV radiation. It is relevant to underline that the climate change also has a significant impact on transportation systems, particularly infrastructure.

Impact of Climate Change

Warming air temperature can raise stream and lake temperatures. It can harm aquatic organisms that live in cold water habitats, such as trout. Warmer water can increase the range of non-native fish species, permitting them to move into previously cold water streams.

The population of native fish species often decreases as non-native fish prey on and outcompete them for food. Impacts of climate change on water availability and water quality will affect many sectors, including energy production, infrastructure, human health, agriculture and ecosystems.

Some regions of the United States, particularly the Northwest, use water to produce energy through hydropower. If climate change results in a lower stream flows in areas where hydropower is generated, it will reduce the amount of energy that can be produced.

Changes in the timing of stream flow can also have an impact on the ability to produce hydroelectricity. Lower water flows would also reduce the amount of water available to cool fossil fuel and nuclear power plants.

Climate change impacts on water supply and quality will also affect tourism and recreation. Quality of lakes, streams, coastal beaches and other water bodies which are used for swimming, fishing and other recreational activities can be affected by changes in precipitation, increases in temperature and sea level rise. In addition, winter sport activities that depends on the production of the snow and ice could be limited in the future as its temperature increases.

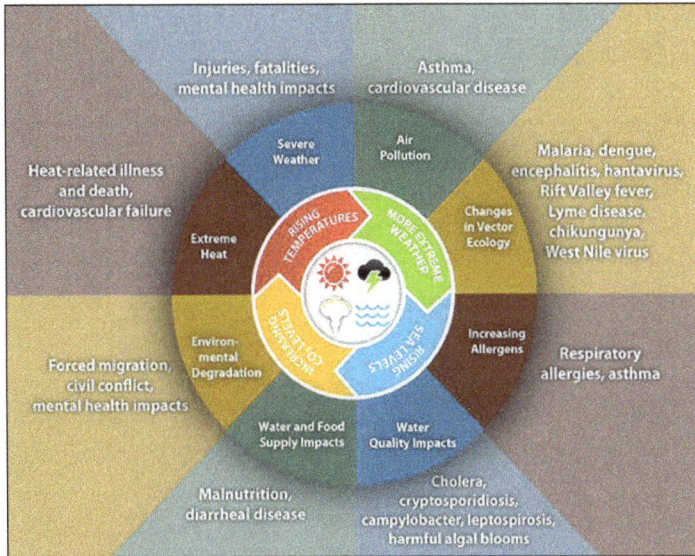

Impact of climate change on human health.

Agriculture and livestock depends on water. Heavy rainfall and the flooding can damage crops and increase soil erosion and delay planting. Areas that experience more frequent droughts will have less water available for crops and livestock. Aquatic species that live in only cold water environments, such as salmon, is affected by rising water temperatures.

Changing water temperatures would also affect the geographic range of fish species. Changes in the availability and quality of water are major concerns for other countries where water resources are already stressed. Planners in many sectors will confront the challenge of the changing water supply. They are likely to adopt variety of adaptation practices designed to better conserve our water supplies, improve water recycling and develop alternative strategies for water management.

Global Warming

The increased input of CO_2 and other greenhouse gases into the atmosphere as a result of human activities will enhance the earth's natural greenhouse effect on raising the average global temperature of the atmosphere near the earth's surface. This enhanced greenhouse effect is called global warming.

Effects of Global Warming

1. Effect of Sea Level

As a result of glacial melting and thermal expansion of the ocean, 20 cm rise in sea level is expected by 2030.

2. Effects on Agriculture and Forestry

High CO_2 level in the atmosphere have long term negative effects on food production and forest growth. More grain belts would become less productive.

3. Effects on Water Resources

Global rainfall patterns will change and the water management strategies of different region will need to adapt to these changes. Drought and flood will become more common while rising temperature will increase domestic water demand.

4. Effects of Terrestrial Ecosystem

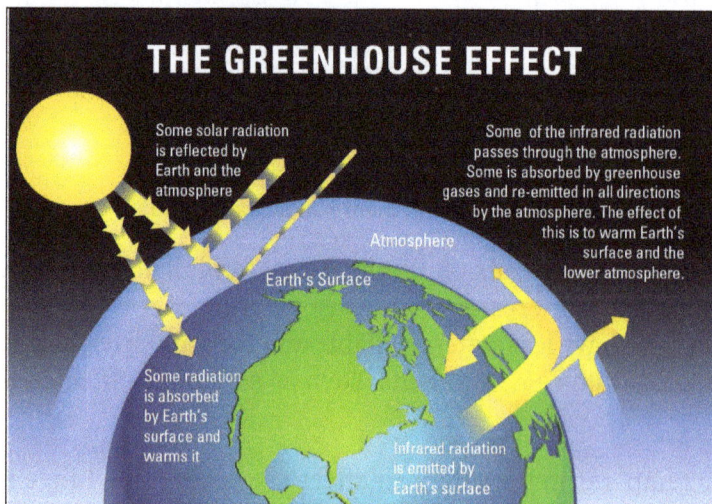

Global warming.

Many plants and animal species will have adapting problems. This will influence the mix of species at different locations. Many will be at risk of extinction whereas, more tolerant varieties will thrive.

5. Effect on Human Health

As the earth becomes warmer, floods and drought become more frequent. There would be increase in water borne diseases, infectious diseases carried by mosquitoes and other disease vectors. The climate change might cause some ecosystem to exceed critical threshold and results in irreversible decline.

Measures to Reduce Global Warming

- Reducing the use of fossil fuels to avoid CO_2 emission.

- Implement energy conservation measures.

- Proper utilization of renewable resources like wind, solar.

- Plant more trees.

- Shift from coal to natural gas.

- Adopt sustainable agriculture.

- Stabilize population growth.

- Efficiently remove CO_2 from smoke stacks.

- Remove atmospheric CO_2 by utilizing photosynthetic algae.

5.5.1 Acid Rain

Acid rain is one of the most dangerous and widespread form of pollution.

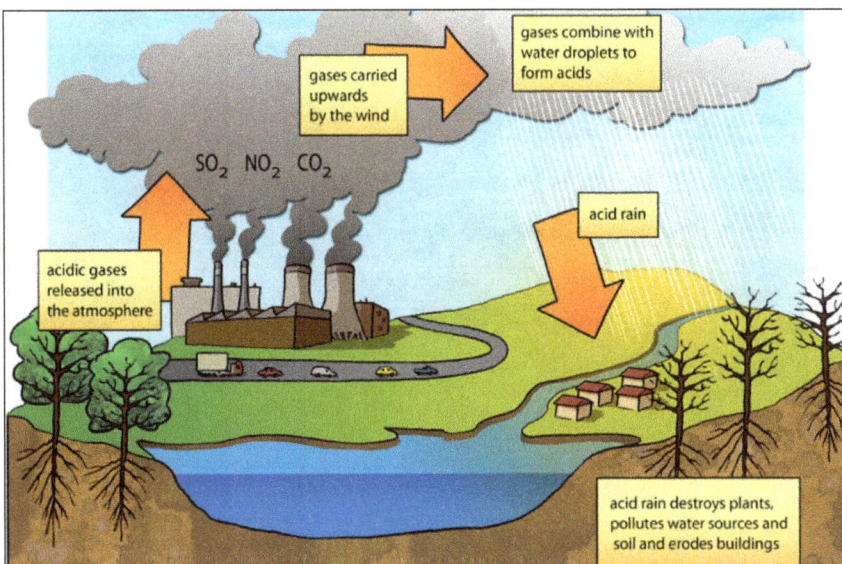

Acid rain cycle.

Sometimes called "the unseen plague", acid rain can go undetected in an area for years. Technically, acid rain is a rain that has larger amount of acid in it than normal.

Acid rain is caused by gases and smoke that are released by factories and cars that run on fossil fuels. When these fuels are burned to produce energy, the sulphur that is present in fuel combines with oxygen and becomes sulphur dioxide. Some of the nitrogen in the air turns into nitrogen oxide. These pollutants go into the atmosphere and becomes acid.

Acid rain is an extremely destructive form of pollution and the environment suffers from its effects. Forests, trees, lakes, animals and plants suffer from acid rain. Humans will become seriously ill and can even die from the effects of acid rain.

One of the major problems that is caused by acid rain is respiratory problems in human being. Many find it difficult to breathe, especially people who have asthma. Asthma, along with dry coughs, headaches and throat irritations can also be caused by the sulphur dioxides and nitrogen oxides from acid rain.

Acid rain can be absorbed by both animals and plants. When the humans eat these plants or animals, toxins in their meals can affect them. Kidney problems, brain damage and Alzheimer's disease have been linked to people eating the "toxic" animals and plants.

Trees are an extremely important natural resource. They regulate local climate, provide timber and forests are homes to wildlife. Acid rain can make trees lose their leaves or needles. The needles and leaves of the trees turn brown and fall off.

Effect of acid rain on forest.

Trees can also suffer from the stunted growth and have damaged bark and leaves, which makes them very much vulnerable to weather, disease and insects. All this happens partly because of direct contact between the trees and acid rain, but it also happens when the trees absorbs soil that has come in contact with acid rain. The soil poisons the trees with toxic substances that acid rain has deposited into it.

Lakes are also damaged by acid rain. A lake polluted by acid rain will support only the hardiest species. Fishes dies off and that removes the main source of food for the birds. Also, birds can die from eating the "toxic" fish and insects and vice versa.

Acid rain can even kill the fish before they are born. Acid rain hits the lakes mostly in the springtime, when fish lay their eggs. The egg comes in contact with the acid and entire generation can be killed. Fish generally die only when the acid level of a lake is high. When the acid level is lower, they become sick, suffer stunted growth or lose their ability to reproduce.

Artwork and architecture can be destroyed by acid rain. Acid particles can land on buildings, causing corrosion. When sulphur pollutants fall on the surfaces of buildings, they react with the minerals in the stone to form a powdery substance that can be washed away by rain. This powdery substance is known as gypsum. Acid rain can damage airplanes, buildings, cars, railroad lines, stained glass, steel bridges and underground pipes.

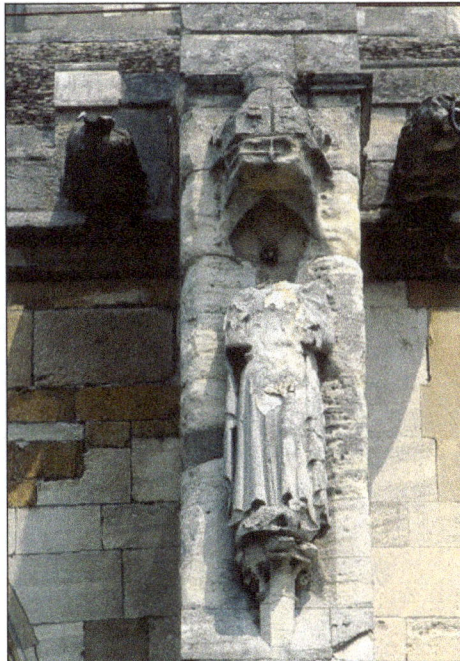

Castle stonework in the UK damaged by acid rain.

The below picture shows how acid rain has eroded the stonework of a castle in Lincolnshire, Currently, both airplane industry and railway industry have to spend a lot of money to repair the corrosive damage done by acid rain. Also, bridges have collapsed in the past due to acid rain.

Stonework of a castle in Lincolnshire, England damaged by acid rain.

5.5.2 Ozone Layer Depletion

An ozone molecule is composed of three atoms of oxygen. Ozone in the upper atmosphere is known as "ozone layer", it protects life on Earth by absorbing most of the ultraviolet radiation emitted by the sun. Exposure to too much UV radiation is linked to cataracts, skin cancer and depression of the immune system and may reduce the productivity of certain crops. Accordingly, stratospheric ozone is termed as "good ozone." In contrast, human industry creates "ozone pollution" at the ground level. This "bad ozone" is a prime component of smog.

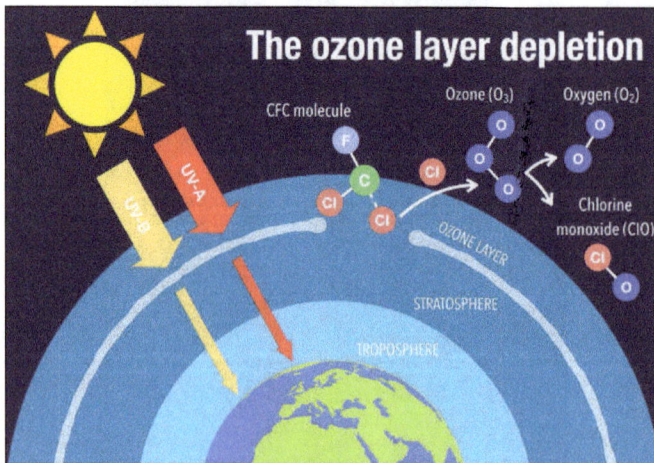

Ozone layer depletion.

The ozone layer is reduced when man-made CFC molecules reach the stratosphere and are broken by short wave energy from the sun.

Free chlorine atoms then break apart molecules of ozone, creating a hole in the ozone layer. The hole in the ozone layer over the Antarctic in 1998 was the largest observed since the annual holes first appeared in the late 1970s.

The CFCs were once used in as foam blowing agents and in aerosol sprays. Their manufacture is now banned by an international treaty, the Montreal Protocol, signed by 160 nations. As CFCs have a long atmospheric lifetime, those manufactured in the year 1970s continue to damage the ozone layer today.

The good news is that scientists predict that the ozone layer will return to its earlier stable size by the middle of the 21st century assuming that nations continue to comply with the treaty.

When the ozone hole was first detected, there was an emotional debate in which many U.S. industries fiercely resisted a ban of CFCs. It took a few years for scientists to show conclusively that human activities was causing the damage. It did not take too long for scientists to invent other chemicals that could replace CFCs for commercial and industrial purposes, but would not harm the ozone layer. CFCs used as propellants were first banned in the United States in 1978.

Process of Ozone Depletion

- CFCs released.

- CFCs rise into ozone layer.

- UV releases CI from CFCs.

- CI destroys ozone.

- Depleted ozone - More UV.

- More UV - More Skin Cancer.

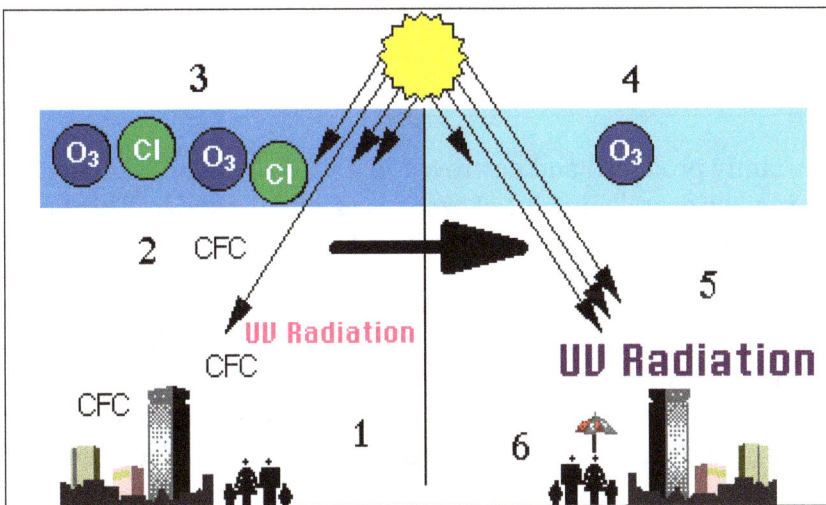

Process of ozone depletion.

The process of ozone depletion begins when CFCs and other Ozone Depleting Substances are emitted into the atmosphere. Winds efficiently mix with the troposphere and evenly distribute the gases. CFCs are extremely stable and they do not dissolve in the rain. After a period of several years, an ODS molecules reach the stratosphere which starts at about 10 kilometers above the Earth's surface.

Strong UV light breaks the ODS molecule. CFCs, HCFCs, carbon tetrachloride, methyl chloroform and other gases release chlorine atoms and halons and methyl bromide release bromine atoms. It is these atoms that destroy ozone, not the intact ODS molecule. It is estimated that about one chlorine atom can destroy about 1,00,000 ozone molecules before it is removed from the stratosphere.

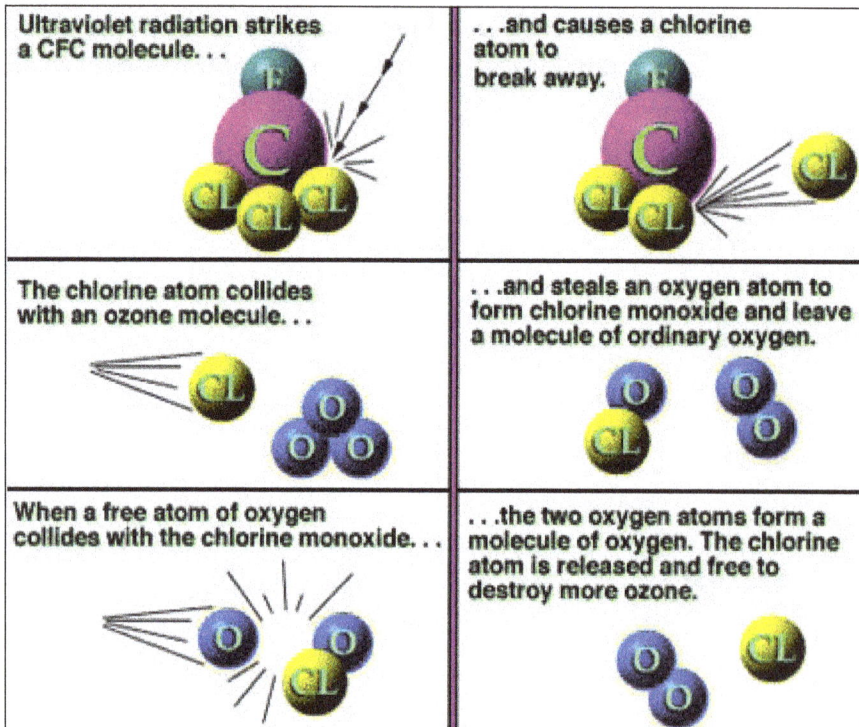

Ozone is constantly produced and destroyed in a natural cycle, as shown in the above figure. However, the overall amount of ozone is essentially stable. This balance can be seen of as a stream's depth at a specific location.

Although individual water molecules are moving past the observer, the total depth remains constant whereas, ozone production and destruction are balanced at ozone levels remains stable. This was the situation until the past several decades that have upset the balance. In this effect, they have added a siphon downstream, removing the ozone faster than the natural ozone creation reactions can keep up. Therefore, ozone levels fall.

Since ozone filters out harmful UltraViolet B radiation, less ozone means higher UVB levels at the surface, larger increase in incoming UVB. UVB has been linked to cataracts, skin cancer, damage to materials like plastics and harm to crops and certain marine organisms. Although some UVB reaches the surface even without ozone depletion, its harmful effects will increase as a result of this problem.

5.5.3 Nuclear Accidents and Holocaust

Nuclear energy can be both beneficial and harmful, depending on the way in which it is used. For example, radiations are used for diagnosis and treatment of cancer, bone defects, blood related diseases in industries. Approximately 17% of the electrical energy generated in the world comes from nuclear power plants.

But in the form of atomic bomb this energy caused devastation in the cities of Hiroshima and Nagasaki. The radioactive wastes from nuclear energy plants and laboratory have caused and continue to cause, serious environmental damage. These are called nuclear hazards.

Chernobyl accident and bombing at Nagasaki are worth mentioning at this place to highlight the devastating power of nuclear energy.

Chernobyl Accident

Although nuclear power has significant benefits, an incident which changed people's attitudes towards nuclear power plants was the Chernobyl Disaster that occurred in 1986.

Chernobyl is a small city in Ukraine, near the border with Belarus, north of Kiev. At 1.00 am on April 25, 1986, a test to measure the amount of electricity that the still-spinning turbine would produce if steam were shut off, was being conducted at the Chernobyl Nuclear Power Station-4.

But the test was delayed and operators failed to program the computer to maintain power at 700 mW and the output dropped to 30 mW. This presented an immediate need to rapidly increase the power and many of the control rods were withdrawn.

In spite of the warning, the operators blocked the automatic reactor shutdown and began the test. As the test continued, the power output of the reactor rose beyond its normal level and continued to rise. The operators activated the emergency system designed to put the control rods back into the reactor and stop the fission.

But it was already too late. In 4.5 seconds, the energy level of the reactor increased two thousand times. The fuel rods ruptured, the cooling water turned into steam and a steam explosion occurred. The lack of cooling water allowed the reactor to explode.

The explosion blew the 1000 metric tons concrete roof from the reactor and the reactor caught fire. This resulted in the world's worst nuclear accident and it took ten days to bring the runaway reaction under control.

Impacts

There were of course immediate fatalities, but the long-term consequences were devastating, followed by evacuation of 1,16,000, of whom 24,000 had received high doses of radiation. Even today many people suffer from illnesses they feel are related to their exposure to the fallout from Chernobyl.

In 1996, ten years after the accident, it was clear that one of the long-term effects was the increased frequency of thyroid cancer in children. There was also a spun in genetic anomalies as doctors began observing clusters of children born displaying monodactyly (fingers fused together to form a paddle) and polydactyly (more than 5 digits on the hands and feet).

Bombing of Hiroshima-Nagasaki

Two beautiful cities of Japan, Hiroshima and Nagasaki were attacked by USA by nuclear bomb near the end of World War II. The nuclear weapon "Little Boy" was dropped on the city of Hiroshima on Monday, August 6, 1945, by the detonation of the "Fat Man" nuclear bomb over Nagasaki followed on August 9. These are to date the only attacks with nuclear weapons in the history of warfare as shown in the figure.

Bombing of Hiroshima-Nagasaki.

The bombs killed as many as 140,000 people in Hiroshima and 80,000 in Nagasaki by the end of 1945, roughly half on the days of the bombings. Amongst these, 15-20% died from injuries or the combined effects of flash bums, trauma and radiation burns, compounded by illness, malnutrition and radiation sickness.

Since then, more have died from leukemia (231 observed) and solid cancers (334 observed) attributed to exposure to radiation released by the bombs. In both cities, the majority of the dead were civilians.

5.6   Wasteland Reclamation

Wasteland

The land which is not in use is called waste land. The waste land is unproductive, unfit for cultivation, graying and other economic uses. About 20% of the geographical area of India is waste land.

Types of Wastelands

Wastes lands can be divided into two types. They are as follows:

- Uncultivated waste lands.

- Cultivable waste lands.

1. Uncultivated Waste Lands

These lands cannot be brought under cultivation.

Example: Barren rocky areas, hilly slopes, stony or leached or gully land or sandy deserts, etc.

2. Cultivable Waste Lands

These are cultivable but not cultivated for more than five year. Cultivable waste lands are important for agricultural purposes.

Examples: Degraded forest lands, gullied water, logged and marsh lands, saline lands, etc.

Causes of Waste Land Formation

- Due to soil erosion, deforestation, over graying, water logging, salinity.

- The increasing demand for firewood and excessive use of pesticides.

- Over exploitation of natural resources.

- By the sewage and industrial wastes.

- Mining activities destroy the forest and cultivable land.

Objectives (or) Need of Waste Land Reclamation

- To improve the physical structure and quality of the soil.

- To prevent soil erosion, flooding and landslides.

- To avoid over exploitation of natural resources.

- To improve the availability of good quality of water for agricultural purposes and industrial operations.

- To conserve the biological resources and natural ecosystem.

- To provide a source of income to the poor.

Methods of Waste Land Reclamation

1. Drainage

Excess water is removed by artificial drainage. This process is used for water logged coil reclamation.

2. Leaching

Leaching is the process of removal of salt from the salt affected soil by applying excess amount of water. Leaching is done by dividing the field in small plots. In continuous leaching 0.5cm to 1.0cm of water is required to remove 90% of soluble salts.

3. Irrigation Practice

High frequency irrigation with controlled amount of water help to maintain better water availability in the Land.

4. Application of Gypsum

Sodicity can be reduced with gypsum. Calcium of gypsum replaced sodium from the exchangeable sites. This process converts clay back into calcium clay.

5. Social Forestry Programme

These programmes induce strip plantation on road, canal sides, degraded forest land, etc.

5.6.1 Consumerism and Waste Products

Consumerism refers to the consumption of resources by the people. It is an organized movement of citizens and government. The special concentration is given to improve the rights and power of the buyers in relation to the sellers.

Consumerism is related to both increase in population as well as increase in our demand due to charge in lifestyle. In the modern society, our needs have increased and so consumerism of resources has also increased.

Traditionally Favorable Rights of Sellers

- The right to introduce any product.

- The right to charge any price.

- The right to spend any amount to promote their product.

- The right to use incentives to promote their products.

Traditional Buyer Rights

- The right to buy or not to buy.

- The right to expect a product to be safe.

- The right to expect the product to perform as claimed.

Important Information's to be known by Buyers

- Ingredients of a product.

- Manufacturing date and expiry date.

- Whether the product has been manufactured against as established law of nature or involved in rights violation.

Objectives of Consumerism

- It improves the rights and power of the buyers.

- It involves making the manufacturer liable for the entire life cycle of a product.

- It forces the manufacturer to reuse and recycle the product after usage.

- The items which are very difficult to decompose like polymeric goods, computers, televisions etc., can be returned to manufacturer for reclaiming useful parts and disposing the rest.

- The reusable packing materials like bottles can be taken back to the manufacturer. It makes the products cheaper and avoids littering and pollution's.

- Active consumerism improves human health and happiness and also it saves resources.

Sources of Wastes

The sources of the waste materials are agriculture, mining, industrial and municipal wastes.

Examples for Waste Products

It includes glass, papers, garbage, plastics, soft drink canes, metals, food wastes, automobile wastes, dead animals, construction and factory wastes.

Effects of Wastes

- The waste released from chemical industries and from explosives are dangerous to human life.

- The dumped wastes degrade soil and make unfit for irrigation.

- E-waste contains more than 1000 chemicals, which are toxic and cause environmental pollution. In computer, lead is present in monitors, cadmium in chips and cathode ray tube, PVC in cables. All these cause cancer and other respiratory problems if inhaled for long periods.

- Plastics are difficult to recycle or incinerate safely because they are nonbiodegradable and their combustion produces several toxic gases.

Factors Affecting Consumerism and Generation of Wastes

1. People Over Population

It occurs when there are more people than the available supply of food and water. Over population causes degradation of resources, poverty and premature death. This situation occurs in less developed countries. Thus in less developed countries per capita consumption of resources and waste generation are less.

2. Consumption Over Population

It occurs when there are less people than the available resources. Due to luxurious life-style per capita consumption of resources is very high. If the consumption is more, the generation of waste is also more and greater is the degradation of environment.

5.7   Environment Laws

Environment Protection Act

This is a general legislation law in order to rectify the gaps and laps in the above Acts. This Acts empowers the central government to fix the standards for quality of air water, soil and noise and to formulate procedures and safe guards for handling of hazard substance.

Environmental Act

- To protect and improve the environment.

- To prevent hazard to all living creatures and property.

- To maintain harmonious relationship between humans and their environment.

Important Features of Environment Act

- The Act further empowers the Government to lay down procedures and safe guards for the prevention of accidents which cause pollution and remedial measures if an accident occurs.

- The Government has the authority to close or prohibit or regulate any industry or its operation, if the violation for the provisions of the act occurs.

- If the violation continuous, an additional fine of rupees five thousands per day may be improved for the entire period of violation of rules.

- The act fixes the liability of the offense punishable under act on the person who is directly in change.

Air Act

AIR (Prevention and Control of Pollution) ACT, 1981

This Act was enacted in the conference held at Stockholm in 1972. It deals with the problems relating to air pollution. It envisages the establishment of central and state control boards endowed with absolute powers to monitor air Quality.

Objectives of Air Act

- To prevent, control and abatement of air pollution.

- To maintain the Quality of air.

- To establish a board for the prevention and control of air pollution.

Important Features of Air Act

- The central board may lay down the standards.

- The central board co-ordinates and settle disputes between state boards, in addition to providing technical assistance.

- The state boards are empowered to lay down the standards for emissions of air pollutants from industrial units or automobiles.

- The state boards are to collect and disseminate information related to air pollution.

- The state boards are to examine the manufacturing orders.

- The state board can adhesive the state government to declare certain heavily polluted areas as pollution control areas.

- The directions of the central board are mandatory on state boards.

- The operation of an industrial unit is prohibited in a heavily polluted areas.

- Violation of law is punishable with imprisonment for a term which may extend to three months.

Water Act

The Water Act was enacted by the Parliament, 1974 for the prevention and control of water pollution and the maintaining or restoring of wholesomeness of water.

The relevant provisions of this act are given below:

- Officials of pollution control board can take samples of water effluent from any industry, stream or well or sewage sample for the purpose of analysis.

- Officials of the state boards can enter any premises for the purpose of examining any record, plant, register or any of the functions of the Board entrusted.

- No person shall discharge any poisonous, noxious or any polluting matter into any stream or well or sewer or on land.

Establish or take any step to establish any industry, operation or process or any treatment and disposal system for any extension or addition there to, which is likely to discharge sewage or trade effluent into a stream or well or sewer or on land.

Bring into use any new or altered outlet for the discharge of sewage.

Begin to make any new discharge of sewage.

- The state board may grant consent to the industry after satisfying itself on the pollution control measures taken by the unit or refuse such consent for reasons to be recorded in writing.

- Any person aggrieved by the order made by the State Board may within thirty days from the date on which the order is communicated, prefer an appeal to such authority.

- The State Board can direct any person who is likely to cause the pollution of water in the street or well to desist from taking such action as is likely to cause its pollution or to remove such matters as specified by the Board through court.

- Pollution control board can issue any directions to any person, officer or authority and such person, officer or authority shall be bound to comply with such directions. The directions include the power to direct.

The closure, prohibition of any industry.

Stoppage or regulations of supply of electricity, water or any other services.

- If any who has been convicted of any offense is again found guilty of an offense involving a contravention of the same provision shall be on the second and on every subsequent conviction be punishable with imprisonment for a term which shall not be less than two years but which may extend to seven years with fine.

- Whoever contravenes any of the provisions of this act or fails to comply with any order given under this act for which no penalty has been elsewhere provided in this act, shall be punishable with imprisonment which may extend to 3 months or with fine which may extend to Rs.10,000 or with both.

Wildlife Protection Act

Protective Act

This act was amended in 1983, 1986 and 1991.

This act is aimed to protect and preserve wildlife. Wild life refers to all animals and plants that are not domesticated. India has rich wildlife heritage. It has 350 species of mammals, 1200 species of birds and about 20,000 known species of insects. Some of them are listed as 'endangered species' in the wildlife (Protection) act.

Wildlife is the integral part of our ecology and plays an essential role in its functioning. The wildlife is declining due to human actions, the wildlife products skins, furs, feathers, ivory, etc., have decimated the populations of many species.

Wildlife populations are regularly monitored and management strategies formulated to protect them.

Important Features

- The Act covers the rights and non-rights of forest dwellers.

- It provides restricted grazing in sanctuaries but prohibits in national park.

- It also prohibits the collection of non-timer forest.

- The rights of forest dwellers recognized by the Forest Policy of 1988 are taken away by the Amended Wild life Act of 1991.

Forest Conservation Act

Conservation Act

This act provides conservation of forests and related aspects. This act also covers all type of forests including reserved forests, protected forests and any forested land. This Act is enacted in 1980. It aims at to arrest deforestation.

Important Features of Forest Act

- The reserved forests shall not be diverted or de reserved without the prior permission of the central government.

- This land that has been notified or registered or forest land may not be caused for non-forest purposes.

- Any illegal non-forest activity within a forest area can be immediately stopped under Act.

Important Features of Amendment Act of 1988

- Forest departments are forbidden to assign any forest land by way of lease or otherwise to any private person or non-government body for reafforestation.

- Clearance of any forest land of naturally grown trees for the purpose of reafforestation is forbidden.

- The diversion of forest land for non-forest uses is cognizable offense and any one who violates the law is punishable.

5.8   Issues Involved in Enforcement of Environmental Legislation

The aim of environmental legislation is to protect the environment, including the health of the people and earth's resources. The enactment of environmental legislation does not mean that the problems are solved.

Once legislation is made at global, national or state level, it has to be implemented. This is possible only by having an effective agency that can collect all relevant data, process it and pass it on to the law enforcement agency.

The people who do not follow the rules must be punished by a legal process. In case cognizance is not given, individuals can file a Public Interest Litigation (PIL) to protect the environment.

NGOs can also take these matters to court in the interest of conserving the environment. A number of legal experts (like MC Mehta) have successfully defended such cases in the court of law.

One of the major issues involved in such cases is illegal gratification in order to get the necessary clearance from the enforcement agency. The public should act as a watchdog so that the concerned authorities are informed and offenders suitably punished.

Three issues that are especially important for environmental legislation are as follows:

1. Freedom of Information

Environmental management and planning is hindered if the public, NGOs or even official bodies are unable to get the information.

A number of laws have been enforced for safeguarding environmental quality. Although, these laws and acts could not be enacted successfully.

2. The Polluter Pays Principle

The polluter pays for the damaged caused by a development this principle also implies that a polluter pays for monitoring and policing. A problem with this type of approach is that fines may bankrupt small businesses, yet be low enough for the large company to write them off as an occasional overhead, that does little for pollution control.

3. The Precautionary Principle

This principle has evolved to deal with the risks and uncertainties faced by the environmental management. The principle implies that prevention is worth a pound of cure, it does not prevent problems but may reduce their occurrence and helps ensure contingency plans are made.

5.8.1 Public Awareness

It is evident that the growing number of poor people, in developing countries due to the rapid population growth with economic constraint which contributes to the degradation of environment and the renewable to the degradation of environment and the renewable sources like forests, water and extinction of various species on which man depends.

For these, greater awareness is needed. Care is necessary to harness the natural resources, so that the quality of the environment does not deteriorate.

One of the reasons for this is improper implementation of the various environmental laws and standards. The most important reason is due to lack of awareness and understanding the implicate environmental degradation.

No governmental programmes more particularly measures to protect the environment can become successful without creating public awareness and enlisting the co-operation. In order to conserve our environment, each and every one must be aware about our environmental problems. To achieve a 'pollution free environment' and have a protected 'green earth', there should be spontaneous coop-eration from the public.

Objectives

The main objectives in creating public awareness regarding the environment are:

- Every citizen to be aware and made aware of importance of environment.

- Reject which are harmful and accept eco-friendly ones.

- Discourage terrorism and report such activities to avoid damage to the ecosystem.

- To conduct meetings, group discussion on development, tree plantation programmes, exhibitions etc.

- To focus on current environmental problems and adopt appropriate ways to solve existing environmental problems.

- To train our planners, decision makers, politicians and administrators.

- To remove poverty by providing employment.

- To take appropriate decisions regarding the use of natural resources.

- To conserve nature and natural resources.

Methods to Create Environmental Awareness

Awareness must be created by both formal and informal education to all sections of the society. Various stages that are useful for raising environmental awareness are given below:

1. Among Students Through Education

Environmental education must be imparted to students' right from childhood stage. It is the most successful method for propagating environmental awareness. Following the directives of Supreme Court, environmental studies are introduced as a subject at all stages including school and college level. This study enable to spreads awareness regarding the protection of the environment.

2. Through Mass Media

Mass media such as Newspapers, magazines, TV and Radio can play a vital role to educate the environmental issues among the public through articles, environmental rallies, plantation campaigns.

3. Entertainment

Environmental awareness can also be propagated through folk songs, street plays, documentaries and Cinema etc. Film about environmental ethics should be prepared and screened in theatre compulsorily with the relaxation of tax free to attract public.

4. Audio Visual Media

To disseminate the concept of environment, special audio visual and slide shows should be arranged in all public places.

5. Voluntary Organizations

The services of voluntary bodies like NCC, NSS and RRC should be effectively utilized for creating environmental awareness.

6. Traditional Techniques

Rural people are much attracted by folk plays, dramas that are used for spreading environmental messages.

7. Arranging Competitions

Story writing, essay writing and painting competitions on environmental issues should be organized for students. Attractive prizes should be awarded for the best effort.

8. Among Planners Decision Makers and Leaders

It is very important to give necessary orientation and training through workshops and training programmes to all section of society.

9. Non-government Organizations (NGO's)

Voluntary organization can help by advising the government about some local environmental issues and at the same time interacting at the grass root levels. They act as a viable link between the two. They can act as an 'action group' or a 'pressure group'.

They can be very effective in organizing public movements for the protection of environment through creation of awareness. WWF India (World Wide Fund for Nature India), CSE (Centre for Science and Environment) and many others play a vital role in creating environmental awareness. The recent report by CSE on more than permissible limits of pesticides in cola drinks sensitized the people all over the country.

The bells are ringing loud, it is up to the public to wake up and act or else perish sooner.

5.9   Population Growth and Variation among Nations

Population Growth and Environmental Issues

The rapid growth of global population for the past 100 years results from the difference between the rate of birth and death. In 1980, global population was about 1 billion people. It took about 130 years to reach 2 billion. But the population reached to 4 billion within 45 years. Now we have already crossed 6 billion and may have to reach about 10 billion by the year 2050 as per the World Bank calculations.

Causes of Rapid Population Growth

- The rapid population growth is due to decrease in death rate and increase in birth rate.

- The availability of antibiotics, immunization, increased clean water, food production and air decreases the famine related deaths and infant mortality.

- In agricultural based countries, children are required to help parents in the fields that is why population increases in the developing countries.

Characteristics of Population Growth

1. Exponential Growth

Now population growth occurs exponentially like 10, 10^2, 10^3, 10^4, etc., which shows the dramatic increase in global population in the past 160 years.

2. Doubling Time

It is the time required for a population to double its size at a constant annual rate. If a nation has 2% annual growth, its population will double in next 35 years.

3. Infant Mortality Rate

It is the percentage of infants died out of those born in the year. Even though this rate has decreased in the last 50 years, the pattern differs widely in developing countries.

4. Total Fertility Rates (TFR)

It is the average number of children delivered by a women in her lifetime. The TFR value varies from 2 in developed countries to 4.7 developing countries.

5. Replacement Level

Two parents bearing two children will then be replaced by their off spring. Due to infant mortality their replacement level is changed. But, due to high infant mortality replacement level is generally high in developing countries.

6. Male - Female Ratio

The ratio of girls and boys should be fairly balanced in a society to flourish. But the ratio has been upset in many countries including China and India. In China, the ratio of girls and boys is 100:140.

7. Demographic Transition

Population growth is generally related to economic development. The death rates and birth rates fall due to improved living condition. This results in low population growth. This phenomenon is referred to as demographic transition.

Problems of Population Growth

- Increasing demands for the food and natural resources.
- Inadequate processing and health services.
- Loss of agricultural lands.
- Unemployment and social political unrest.
- Environmental pollution.

Population Growth Variations

From the dawn of mankind to the turn of the 19th century world population grew to a total of one billion people. During the year 1800s, human numbers increased at increasingly higher rates, reaching a total of about 1.7 billion people by the year 1900. World population has grown even more rapidly during the present century, with the greatest gains occurring in the post-World War II period and stands at over three times its size in 1900 (i.e) some 5.9 billion people today.

Population growth has continued throughout the past three decades in spite of the decline in fertility rates that began in many developing countries in late 1970s and the toll taken by HIV/AIDS pandemic.

While the rate of increase is declining, in absolute terms world population growth continues to be substantial. Global population increase is equivalent to adding the population of Egypt, Gaza, Israel and Jordan to the existing world total each year.

According to Census Bureau projections, world population will increase to the level of nearly 8 billion persons by the end of next quarter century and will reach 9.3 billion persons, a number more than half again as large as today's total by 2050.

The future of human population growth has been evaluated and is now largely being decided, in the world's Less Developed Nations. 96% of world population increase now occurs in developing regions of Africa, Asia and Latin America and this percentage will rise over the next quarter century.

90% of the world's births and 77% of its deaths will take place in 1998. Census Bureau's projections indicate that early in the next century, crude death rates will exceed crude birth rates for the world's More Developed Countries and the difference natural increase will be negative.

At this point, international migration will become a critical variable determining whether the total population of today's MDC's increases or decreases. These projection shows negative natural increase offset by net international immigration through 2019 but, if present trends continue, the population of the world's MDC's will slowly begin to decrease from the year 2020 onward.

Variation Among Nations

As the growth rate in world's more affluent nations becomes negative, all of the net annual gain in global population will come from the world's developing countries.

Below is a listing of most populated countries in the world:

- Bangladesh → 147,365,352.

- Brazil → 188,078,227.

- China → 1,313,973,713.

- Germany → 82,422,299.

- India → 1,095,351,995.

- Indonesia → 245,452,739.

- Japan → 127,463,611.

- Mexico → 107,449,525.
- Nigeria → 131,859,731.
- Pakistan → 165,803,560.
- Philippines → 89,468,677.
- Russia → 142,893,540.
- United States → 298,444,215.
- Vietnam → 84,402,966.

5.9.1 Family Welfare Programme

Family welfare programme was implemented by the Government of India as a voluntary programme. It is an integral part of the overall national policy of growth covering maternity, human health, child care, family welfare and woman's right.

Objectives of Family Welfare Programming

- Slowing down the population explosion by reducing the fertility.
- Pressure on the environment due to over exploitation of natural resource.

Population Stabilization Ratio

The ratio is derived by dividing crude birth rate by crude death rate.

Developed Countries

The stabilization ratio of the developed countries, which is more or less stabilized indicating zero population growth.

Developing Countries

The stabilization ratio of the developing countries nearing 3, is expected to lower down by 2025. Stabilization in developing countries is possible only through various family welfare programme.

National Family Welfare Programme

Previously this programme was known as National Family Planning Programme. In the year 1977 the name was changed to National Family Welfare Programme. Family planning programme was launched in India in 1952. India was the first country to do so.

Beginning of the programme was modest. i.e. establishment of few FP clinics distribution of FP educational material, training of health functionaries and research. During the third 5-year-plan (1961-66) family planning was declared as centre of planned development.

Then the emphasis was shifted from clinic approach to extensive education approach (i.e.. motivating people about small family norm). A separate Department of Family Planning was created in 1966 in the Ministry of Health. In 1972, the MTP Act was passed. In April 1976, National Population Policy was framed.

During the emergency period (1976), forcible sterilization campaign led to the defeat of Congress in 1977 elections. In June 1977, new Janata Government formulated a new population policy and made family planning as voluntary and renamed it as Family Ware Programme.

The acceptance of primary health care approach as the key to the achievement of health for all by 2000 AD led to the formulation of National Family Welfare Programme in 1982.

Importance of Family Welfare Programme

- The family welfare programme occupies an important position in the nation's socioeconomic development.

- Indian population which was 14 crores in 1947 has crossed 100 crore mark by 2000 AD. India has only 2.4% of world's land area but it supports about 15.5% of world's population.

- India's population is increasing by 1.8 crores every year. To check this galloping growth, the country has laid down long-term demographic goal of achieving an NRR of one by the year 2000 AD.

- Acceptance of the family welfare services is made voluntary.

- The programme was 100% centrally sponsored scheme. FP programme was integrated with the MCH services.

Organizational Set Up

1. Central Level

At central level Central Cabinet Subcommittee is present. It is headed by Prime Minister. Next level is Population Advisory Council. This is headed by Union Minister of Health and Family Welfare. Members are representatives of various professional bodies and some technical persons. Next level is Central Family Welfare Council, which is headed by union minister and ministers of health and family welfare of all states. It coordinates the work of the programme.

National Institute of Health and Family Welfare, situated in Delhi, is the apex institute. It undertakes research and training in family welfare.

Directorate General of Health Services was the central programme officer for Family Planning. He advises Government of India on various aspects of family welfare.

2. State Level

Ministry of Health and Family Welfare is the apex organization at the state level. This is headed by the minister of health and family welfare of the respective state. At the state level the family welfare work is organized by State Family Welfare Bureau.

The State Family Welfare Bureau has three wings:

- Administrative wing (headed by state family welfare officer and associated by some officers),

- Education and information wing (headed by mass media information officer),

- Field operation and evaluation wing (headed by statistical officer).

3. District Level

At district level the work of family welfare is organized by District Family welfare Bureau. This has three wings like the state level. At some districts Regional Family Online Training Centres are present. These will undertake training of medical officers and para-medical staff.

4. Peripheral Level

In rural areas the family welfare work is looked after by rural family welfare centres attached to PHC while in urban areas urban family welfare centres will took after this work.

5. Village Level

At village level the MPHA (F) and MPHA (M) are mainly responsible for the programme. They will take the assistance of CHG, TBA and anganwadi workers.

Goals of National Population Policy

- NRR I (which implies two-child norm).

- Birth rate 21/100 population.

- Death rate 9 per 1000 population.

- Raising couple protection rate to 60%.

- Reduction of family stet to 2.3.

- Decrease the IMR to 60 per 1000 live births.

Programme Strategies

- Integrated approach.

- Cafeteria approach.

- Welfare approach.

- At risk approach.

1. Integrated Approach

It is the integration of all the activities like raising the age of marriage. increasing female literacy, empowerment of women. raising the overall economy of the country etc. To achieve these objectives, various anti-poverty programmes have been started.

2. Cafeteria Approach

The programme is made voluntary and all pressurizations are removed. It stresses more on motivation and education of the people about the benefits of family welfare programme.

3. Welfare Approach

In 1977, the word planning was removed and changed to family welfare programme. This indicates the government's commitment to family welfare rather than family planning.

4. At risk Approach

Like any other programme NFWP stresses to ensure maximum support to those who are in need.

Now family welfare has been given priority in the health development. The Government of India has evolved a comprehensive population policy.

Permissions

All chapters in this book are published with permission under the Creative Commons Attribution Share Alike License or equivalent. Every chapter published in this book has been scrutinized by our experts. Their significance has been extensively debated. The topics covered herein carry significant information for a comprehensive understanding. They may even be implemented as practical applications or may be referred to as a beginning point for further studies.

We would like to thank the editorial team for lending their expertise to make the book truly unique. They have played a crucial role in the development of this book. Without their invaluable contributions this book wouldn't have been possible. They have made vital efforts to compile up to date information on the varied aspects of this subject to make this book a valuable addition to the collection of many professionals and students.

This book was conceptualized with the vision of imparting up-to-date and integrated information in this field. To ensure the same, a matchless editorial board was set up. Every individual on the board went through rigorous rounds of assessment to prove their worth. After which they invested a large part of their time researching and compiling the most relevant data for our readers.

The editorial board has been involved in producing this book since its inception. They have spent rigorous hours researching and exploring the diverse topics which have resulted in the successful publishing of this book. They have passed on their knowledge of decades through this book. To expedite this challenging task, the publisher supported the team at every step. A small team of assistant editors was also appointed to further simplify the editing procedure and attain best results for the readers.

Apart from the editorial board, the designing team has also invested a significant amount of their time in understanding the subject and creating the most relevant covers. They scrutinized every image to scout for the most suitable representation of the subject and create an appropriate cover for the book.

The publishing team has been an ardent support to the editorial, designing and production team. Their endless efforts to recruit the best for this project, has resulted in the accomplishment of this book. They are a veteran in the field of academics and their pool of knowledge is as vast as their experience in printing. Their expertise and guidance has proved useful at every step. Their uncompromising quality standards have made this book an exceptional effort. Their encouragement from time to time has been an inspiration for everyone.

The publisher and the editorial board hope that this book will prove to be a valuable piece of knowledge for students, practitioners and scholars across the globe.

Index